Evaluating the Science and Ethics of Research on Humans

Evaluating the Science and Ethics of Research on Humans A Guide for IRB Members

Dennis J. Mazur, M.D., Ph.D.

Chairman, Multisite Institutional Review Board
Northwest Health Network, Department of Veterans Affairs
and
Professor of Medicine and Senior Scholar
Center for Ethics in Health Care
Oregon Health and Science University
Portland, Oregon

The Johns Hopkins University Press / Baltimore

Printed in the United States of America on acid-free paper
9 8 7 6 5 4 3 2 1

The Johns Hopkins University Press
2715 North Charles Street
Baltimore, Maryland 21218-4363
www.press.jhu.edu

Library of Congress Cataloging-in-Publication Data
Mazur, Dennis John.
Evaluating the science and ethics of research on humans : a guide for IRB members / Dennis J. Mazur.
p. ; cm.
Includes bibliographical references and index.
ISBN 0-8018-8501-9 (hardcover : alk. paper)—ISBN 0-8018-8502-7 (pbk. : alk. paper)
1. Human experimentation in medicine—Moral and ethical aspects. 2. Medical ethics. 3. Medicine—Research—Moral and ethical aspects. I. Title.
[DNLM: 1. Ethics Committees, Research—organization & administration—United States. 2. Human Experimentation—ethics—United States. 3. Ethical Review—United States. 4. Human Experimentation—standards—United States. W 20.55.H9 M476e 2006]
R853.H8M39 2006
174.2'8—dc22
2006012326

A catalog record for this book is available from the British Library.

To my lovely wife, Pantipa, and my sons, Marcus and Matthew, for their support and encouragement over many years of IRB work

To my mother and father, Irene and Joseph; my aunt and uncle, Stephanie and Louis; my brother, Larry; and Hutch, Cindy, Joe, and Tony

Contents

Preface

The purpose of this book is to help members of institutional review boards (IRBs) better understand their role in the important task of judging whether research projects submitted to them for review should be allowed to proceed. The book takes the board member through the questions he or she must ask, the types of decision making that will be encountered during the review process, and the approaches that will be required to perform the IRB's primary task—protecting human participants in research.

This book concentrates on the key areas of research on humans: drugs, medical devices, survey research, genetic information, and behavioral research. It also addresses research involving exposure of humans to substances to determine toxicity. The reader should understand, however, that the issues raised here are in no sense exhaustive of the problems IRBs confront, because the research questions that scientists generate change over time.

The approaches and information presented herein come out of my thirty years of research in informed consent and fifteen years serving as chair of the IRB of the Department of Veterans Affairs Medical Center in Portland, Oregon. I came to my role as chair of an IRB with a somewhat nontraditional research background in cognitive psychology, ethics and the law, and risk communication. However, my focus has been on rational decision making in the patient-physician relationship. I helped develop a multisite regional IRB, and the perspectives gained through that experience are shared here. This book presents methods that I have found useful in training IRB members to identify ways in which study participants can be protected and to decide whether a research proposal as submitted will do enough to protect the participants. My approach is based on identifying key problem areas that the individual IRB members can recognize, consider, and then bring to the full board for discussion and decision making.

Involvement of an IRB may start early, in the hypothesis-generation or hypothesis-refinement process of a study, or it may start with the review of a scientific protocol and an informed consent form that were developed without IRB input. The IRB processes involve discussions with principal investigators and extend to meetings with study sponsors and medical product manufacturers. Often, there are multiple exchanges of information and discussions, with the IRB asking a host of questions and receiving a host of answers, and inevitably some answers generate new questions. This process continues until the IRB makes a decision to approve or reject the proposal.

Some people argue that the "science" of a research study should be reviewed by a research service and only the "ethics" of that research study should be reviewed by an IRB. However, many of the ethical questions an IRB needs to pose cannot be adequately formulated or even addressed until the board understands the science behind the study's hypothesis. The understanding of the science of a research protocol then progresses—through its development in the scientific methods, the weighing of the risks of the scientific endeavor, the selection of individual participants for inclusion in the study, close observation of all laboratory and study data being collected and quick action on any abnormalities detected, and the analysis of the data that the study has secured at the conclusion, and even beyond. The scientific and statistical issues may be complex, and the IRB may need to solicit the opinions of local, regional, or national experts.

Research involving human participants has taken some great strides since its beginnings. The ethical principles of nonmaleficence, beneficence, autonomy, and quality of life are now well understood as basic concepts in clinical care. Beyond that, however, the concepts of communication and understanding in research involving human participants need to be understood by all: researchers, study participants, IRB members, and anyone engaged in or interacting with the research enterprise. For example, IRBs help study sponsors and principal investigators make clear to study participants the distinction between clinical care (providing optimal care of the patient) and clinical research (attempting to generate knowledge to optimize care for future populations).

Explaining this distinction is one of the tasks of the informed consent form, but even after participants have digested the informed consent form and discussed it with the principal investigator, they may have difficulty distinguishing between clinical care and clinical research. Does an individual with an incurable disease who has no clinical care options but is offered participation in a study of a new drug truly hear that distinction clearly?[1] How should the study sponsor and principal investigator best remind study participants, in an ongoing fashion through the study, that they are involved in research if the study is being conducted in the same medical center with the same physicians and teams who deliver the patient's clinical care? Identifying such difficulties in communications between the principal investigator and the study participants and attempting to eliminate them is just one part of the IRB's task.

As it systematically reviews the scietific protocol and informed consent form of each research study that comes before it, an IRB must consider questions like: What happens when those in a medical institution fail to distinguish between the work of the IRB and the work of the research service? Which group is responsible for primary review of the science? When does that primary review occur and to what depth? What happens when groups disagree on what research studies an institution can safely take on within its current clinical care responsibilities? Who is to decide when, if ever, an institution can take on more risky research studies involving human participants?

I hope that this book will assist IRB members, especially new members, in the tasks and the challenges they encounter in their work on the IRB and in their interactions with principal investigators and the rest of their institution. New members will be encouraged to learn that often the insight of new members illuminates the workings of the IRB as it continually reexamines its primary focus: the optimal protection of the participants of a research study.

I am grateful to the following colleagues for their support over time: Wayne Clark, Sola Whitehead, Angie Lacey, Lisa Gunion-Rinker, Vickie Vonderoe, Stephen Hefeneider, Richard Jones, Susan Hart, James Reuler, Gordon Noel, David Hickam, Mark Helfand, Bev Jefferson, Ted Galey, John Kendall, Leslie Burger, and Carol Frank. I thank the members of the Portland VAMC Institutional Review Board and the members of the VISN 20 Multi-Site Institutional Review Board. I am also grateful for the support and encouragement of Richard Jones, Ursula Helmut, Marcus Mazur, Max Metcalf, Bill Wickham, Lawrence Oresick, Mabel Gearhart, and Thomas E. Peterson.

Thanks go as well to all of the regulators it has been my privilege to work with over the years. I advise anyone who has questions about research regulations to ask the appropriate regulator with expertise in interpreting the regulations.

I also thank Wendy Harris, of the Johns Hopkins University Press, for her tremendous encouragement, help, and support through the completion of this project; and I thank Anne Whitmore, of the Press, for help in clarifying the manuscript.

Evaluating the Science and Ethics of Research on Humans

Introduction

What Can the New IRB Member Expect?

As they reflect on the letter of appointment to serve on an IRB, new members may have a number of questions. Questions will come to mind as they think about attending their first IRB meeting and what will be expected of them. Questions will come even faster as they receive the first packet of papers from the IRB secretary or a member of the research staff. It is to be hoped that the packet will arrive at least one week before the meeting, but, given the pressures of research and funding submission deadlines, there are no guarantees.

Who Appoints IRB Members and How Are They Selected?

Who appoints a person to an IRB depends on the institution in which the IRB resides or has jurisdiction over, which may be a medical institution that conducts research studies or a pure research institution that does not offer medical care. The appointment of new IRB members is most typically done by the institution. If there is a research service with a head of research, that individual will probably make the appointments. The research service is the administrative department in charge of and ultimately responsible for the research that goes on within that institution.

There are many reasons for being selected to serve on an IRB. The institution's standard operating procedures will determine the rotation of people on and off the IRB. In addition to the voting members of the IRB, there may be ex officio members who attend meetings only as needed.

Certain IRBs may want to have on the committee people with specialized areas of expertise. With so many tasks, it can be difficult for a particular IRB to have as much expertise as it needs to do an optimal job of reviewing proposed research. IRB members trained in conducting systematic searches of the peer-reviewed medical literature are valuable, as they can assure that this literature has been searched appropriately and in depth whenever the need arises, and the need will arise often.

Similarly, an IRB member who is a scientist with a background in the development, review, and evaluation of scientific protocols is valuable. A sound methodologist can examine and explain what is going on in a particular scientific protocol, translate the science for the nonscientist members of the IRB, and review a study's methods.

An IRB member who is an editor can be a valuable asset to an IRB that must make recommendations on how best to translate the difficult wording of an informed consent form into language an individual who is not a scientist can understand.

A background in the assessment of competence and decision-making capacity is another valuable asset. Such knowledge is rare, however, because research on assessing decision-making capacity is still in its infancy. Thus, although experts may have suggestions, the question of how best to assess decision-making capacity remains unanswered.

Another compelling skill is accurate recall. People with this capacity, after they have served on an IRB for a while, can provide history of the IRB's decision making over time. They can remind the board how it previously handled particular situations, then the board can discuss whether the issue in front of it should be handled as in the past or a new approach should be used.

There is often a battle within an IRB or between an IRB and the research department over the sometimes competing goals of protection of human study participants and the development of new research within the institution. There may be a battle in the selection criteria for IRB membership, the qualifications of the candidate, or the nature of the candidate's values. An IRB member can be viewed by the institution as a tool that is more or less useful, given the type of candidate the institution desires as a new IRB member. If there are problems in the protection of research subjects or with getting new research protocols through the current IRB, these problems may influence the selection criteria for IRB membership.

Over the course of an institution's IRB history, various motivations may direct changes in the membership on an IRB. Some apply generally:

- All principal investigators should rotate through the IRB, to learn how to optimally protect research participants.
- Keeping key people on the IRB will ensure a consistency to the process of protecting participants.
- The IRB needs fresh ideas.

Some motivations are more pointed and potentially problematic:

- When a current member of the IRB is seen as overprotecting research participants at the expense of research progress, this individual may be targeted for early exit from the IRB.
- When an IRB is viewed as overprotective, the institution may seek a new member with a dominating presence or boisterous presentation style to

lead the IRB and direct IRB attention one way or another on issues that are currently dividing the IRB and slowing down the IRB's review of scientific protocols and informed consent forms.

- If a recent review by a regulatory body has scolded the institution for its lack of protection of vulnerable individuals in research, the institution will need to shift the balance of the IRB toward increased protection of participants.
- If a recent review by a regulatory body has praised the institution for the IRB's protection of vulnerable study participants, the institution may want the IRB to take on projects of a more controversial nature for future research at the institution.

The individuals who control who serves on an IRB thereby influence the IRB and its decision making. Changes in membership can be designed to focus the IRB's attention away from the protection of participants toward the facilitation of research. For example, a newly appointed IRB member may bring instructions from a research service chief or a research committee whose goals are to further the research interests of study sponsors, principal investigators, and others instead of fostering the protection of participants. Or an IRB member may discourage other members from requiring clarification of terms and concepts in informed consent forms, thus allowing the use of documents potentially confusing to the individual considering study participation.

The battle over the IRB and its members and their selection is always the same: the attempt to shift the IRB's attention away from the primary goal of protecting human participants toward other goals, most notably the facilitation of research. The focus of many parties trying to facilitate scientific research is not to acquire new general scientific knowledge to benefit future generations but to acquire the financial rewards that research endeavors may generate through business arrangements, relationships, and enterprises. This is where conflicts of interest arise and need special IRB attention.

Subcommittees

The IRB is defined as a "board committee," so it is sometimes referred to as "the IRB committee." It usually does some of its work in subcommittees. Most IRB members sit on subcommittees, which are made up of IRB members and other experts, called in during review of a proposal to clarify a particular topic area. An IRB subcommittee reports its findings back to the full IRB. Subcommittees are not to conduct independent decision making. The strength of the IRB is in decision making by the full IRB with all members present.

Subcommittees of the IRB may deal with any topic that needs further review and study. General topics include: what certain regulations mean, how certain regulations apply to a particular issue, and how decision-making capacity is to be assessed in the research studies being conducted by the institution's medical, psychiatric, and surgical services. Subcommittees can also be assigned more specific topics, such as how the institution's emergency room could be restructured to conduct certain types of studies, whether the institution should allow placebo-controlled drug studies in its research program or only studies of a new drug versus a standard drug, and what limits will be placed on the type of placebo-controlled study the institution will allow.

If the subcommittee has done its work, the IRB's overall decision making will be improved. Although a well-selected subcommittee can be of significant help in a particular area, it cannot take over the role of an IRB.

For How Long Do IRB Members Serve?

The institution's standard operating procedures will specify the length of an IRB member's term. Some members wish to serve only a minimum term; others may wish to serve continuously or as long as the procedures allow.

How Much Time Will IRB Work Take?

The number of researchers in an institution requesting to conduct research on humans is a measure of how much work an IRB will have. Some members will spend days on a scientific protocol and informed consent form assigned to them, while others will do a minimal amount of work and depend on more experienced members to fill the gaps before any vote is taken.

An institution may have more than one IRB, each with its own chair, alternative chair, and set of members, and the approaches of the IRBs may diverge markedly. It is problematic when principal investigators begin to view one IRB as less thorough, less systematic, or less demanding than another IRB, because a principal investigator may try to submit a study to the less-demanding IRB. In my opinion, an institution with more than one IRB needs to have an overseer of IRBs whose job is to identify and alert the chair of each IRB to these differences.

Is IRB Work Constant or Seasonal?

Because grant submission deadlines often cluster during the year, the IRB can expect that its work will often increase at certain times. Grant submission deadlines are times of high anxiety for principal investigators, the administrative staff of a research service, and the IRB if the funding agency requires that an IRB review the grant before it is submitted.

What Will Happen before a New Member's First IRB Meeting?

Before the first IRB meeting, the new member will have been trained on how to review research involving human participants. Each institution will base its training on the federal regulations under which an IRB functions. The member will have received some training regarding the U.S. *Code of Federal Regulations* and the Belmont Report or other foundational documents (to be discussed in later chapters). Along with copies of the relevant sections of the *Code of Federal Regulations,* the member will have received the institution's manual of standard operating procedures.

Before the full IRB meeting, members typically receive prepared materials, including an agenda, the minutes of the previous meeting, a list of all studies to be reviewed, the names of each study's principal investigator and research staff, and an assignment of research studies. Members may receive information about the qualifications of the principal investigator and research staff and the abilities of each member of the research team to conduct an aspect of the research study. For the first meeting, an IRB member should go through all of the materials provided, to get a feel for the types of issues the IRB will discuss at this meeting. A new IRB member may want just to sit in on the first few meetings, to get a better understanding of how the IRB functions.

A new IRB member may be paired with a more senior member to learn how to conduct a review and discussion related to a study hypothesis, scientific protocol, and informed consent form. A new IRB member may have fresh insights on how the IRB is functioning and should write down all questions or comments that he or she has on how the IRB is working, for discussion with the chairperson or other IRB members at the end of the meeting.

Who Sets the Agenda?

The agenda may be developed by the research staff, working with the chair of the IRB (or the chairs of the institution's IRBs). Often, an experienced chair can recognize potential trouble areas and will assign a difficult scientific protocol to a senior member. Sometimes the head of the research service may have heard concerns or complaints from principal investigators and may want these issues discussed at the meeting of the full IRB.

Does the IRB Chair Always Run the Meeting?

The IRB chair usually runs the meeting. Each IRB should have an alternative chair, to take over when the chair is not present or when the chair has a conflict of interest with a research study being reviewed.

What Can New IRB Members Expect at Their First Meeting?

As the new IRB member attends a first meeting, he or she will hear the banter that occurs before the start of any meeting. The new member can expect to see quite a diverse membership if the IRB sits in an academic medical center that conducts research studies and trials over a wide range of patient groups. For example, the representation of special populations is often accomplished by having scientists and nonscientists who care for vulnerable individuals or representatives of vulnerable patient groups in attendance or on the IRB as voting members.

The minutes of the previous meeting will be reviewed, to confirm that they accurately reflect the IRB's decisions made at that meeting. It is difficult to capture debates in a set of minutes, so each IRB member should make certain that his or her position is accurately reflected in the minutes and should make amendments to the minutes if neccessary, which will need the full board's approval. IRB members will vote on acceptance of the minutes of the previous meeting either as submitted or as amended.

The IRB may have asked the principal investigators of the study or studies to be discussed to come to the meeting, perhaps to clarify a point in a scientific protocol or to defend the risk to be undertaken in a study. Sometimes a principal investigator will have asked the IRB for time on the agenda, to bring up issues. The IRB can ask questions of the principal investigator and then have the principal investigator leave the room. IRB members' opinions may differ on whether a principal investigator should be invited to a meeting. Some members may want to see the principal investigator's response in writing instead. Even if the principal investigator appears in person, written communication is essential, to avoid misunderstandings regarding the principal investigator's intentions.

What Can New IRB Members Expect of the Others on the Board?

Although every IRB member is trained in the federal regulations, members may have differing opinions on what it means to optimally protect participants. Each member will have a different threshold for when he or she becomes concerned about a scientific protocol or informed consent form and calls the IRB's attention to the way it is approaching a particular issue.

The new IRB member will often find that, when the chair introduces the member, some colleagues will be less collegial than the new member might expect. Sometimes new members are affronted by the seeming rudeness of remarks or by the taking of positions in favor of research over the protection of study participants by other members of the board. In the application of the theory espoused by the *Code of Federal Regulations* and the Belmont Report to the practice of IRB functioning, the discussion of what should and should not be allowed in a research study can often reach a boiling point.

During an active debate, senior IRB members may recall a similar discussion that happened months, or even years, earlier. The meeting minutes will be helpful in refreshing the IRB's collective memory of this decision making, yet a new member may disagree wholeheartedly with the IRB's previous approach.

Some members of the IRB may be trying to persuade the institution to engage in types of studies that the IRB in the past deemed too risky for that institution. For example, suppose the IRB is reviewing a proposal for a psychiatric study to be conducted in the institution's busy emergency room. The principal investigators are proposing that individuals coming to this emergency room in an acute psychotic or schizophrenic state be asked to participate in a research study, without previous discussion with these individuals or their family members. The IRB's primary reviewer immediately poses the question, How can it be assumed that an individual will have the capacity to make the decision to enroll in this study? The members present at the meeting may respond with a variety of questions, such as:

- How else can we understand how best to manage and treat acute psychotic and schizophrenic states?
- How will the research team assess the decision-making capacity of the individual coming to a busy emergency room?
- Can advance directives be developed that would allow individuals in acute mental health states to enroll in a research study?
- What role, if any, would a family member or significant other have in an individual's consent to participate in a research study?
- Should the proposed study be rejected outright?
- How would the institution's emergency room team's procedures and its interface with the in-house team of psychiatrists and the research team have to be modified to allow consideration of future studies with individuals in less-acute mental health states?

This example shows the differing views IRB members may express at a meeting. It also indicates the need to involve the personnel who will be executing the decision if the institution would be modifying its procedures in consideration of the research study.

Are IRB Discussions Confidential?

Even after conclusion of a debate and a vote has been taken, IRB members may hold different perspectives on what the decision making was about and even on what decision was made. Therefore, it is best for an IRB member not to discuss who made what comments during the meeting and how these were interpreted. The meeting minutes are the official representation of IRB decision making, and all questions about decisions should be addressed to the individual who was in charge of the review and the discussion. Often, right after a meeting, a principal investigator will question an IRB member on a particular scientific protocol, while the IRB member may need time to reflect on the discussion, so responding to the inquiry could create misunderstanding. The IRB may want the chair to take on the role of communicating with principal investigators or may specify someone else to take on this often difficult role.

What Can New Members Expect on a Regional or Multiple-Site IRB?

When institutions combine their efforts and form a regional or multiple-site IRB, each institution must have enough IRB members on the IRB to assure optimal protection of study participants at each site. On a regional or multiple-site IRB, a new member may be challenged, so that the other members can make certain that there is enough expertise in the institution to understand the science and ethics of each research study and optimally protect the participants.

The new member of a regional or multiple-site IRB may feel overwhelmed by the tasks of representing his or her institution and evaluating each scientific protocol and informed consent form. For the IRB members to have the support they need, members must be able to conduct appropriate and systematic discussion with the individuals responsible for conducting or overseeing research at their institution. There must be clear communication within each institution regarding the limits of the types of research conducted in the institution. Does the institution have sufficient expertise to conduct the scientific research itself? Does the institution have the abilities to monitor and oversee principal investigators and research staffs and the way research is conducted to optimally protect study participants?

How New IRB Members May React to Their Work and What They Get Out of IRB Work

Although the tasks of the IRB are difficult, they are surmountable. New IRB members should not be overwhelmed by their first view of IRB meetings. They should do the best they can and bear in mind that they bring to the IRB a fresh perspective on the business that the IRB is conducting. New IRB members may be inclined not to speak up at meetings when they do not understand an issue or topic, are not following a discussion, or do not agree with what is being said. But, speaking up is key because when any one IRB member has a question and speaks up, it will often be immediately recognized that other members had the same question or concern. New IRB members can energize an IRB with insights.

The work of an IRB is often most satisfying because of the exposure a member gets to new scientific topics, perhaps at their inception, when the ideas have been formulated but may well need revision from both scientific and ethical standpoints. IRB members help early research that may eventually come to be published. They help investigators and study sponsors by pointing out the best ways they can modify their research proposals to better protect study participants. Being a member of an institutional review board is a critical role, because it is the IRB that has the main oversight of all research within an institution with respect to the protection of human subjects, and without that oversight the research cannot go forward.

Part I

The IRB, Its Work, and Its Challenges

1 What Is an IRB, and What Does It Do?

Medical research is important in our lives on a daily basis. Medical research involving human study participants, or human subjects, is of central importance. (When human subjects of research have been asked how they prefer to be referred to, many have responded that they prefer being called study participants rather than study subjects.) As a result of research, advances are made in all branches of medicine, including surgery and mental health care, and these advances affect individuals and whole populations of people.

Most medical research is done at a university and/or medical center and is funded by both government and private enterprise. The federal government, institutions, study sponsors, and everyone else involved in research need to be concerned for the optimal protection of study participants. Institutional review boards have been put into place to systematically carry out this oversight function.

The U.S. *Code of Federal Regulations* (*CFR*) embodies the general and permanent rules published in the *Federal Register* by the executive departments and agencies of the federal government. Each volume of the code is updated once each calendar year and is issued on a quarterly basis. This document is sponsored by the Office of the Federal Registrar, National Archives and Records Administration, and may be viewed on the U.S. Government Printing Office website (www.access.gpo.gov/nara/cfr).

The *CFR* is divided into fifty titles, which represent broad areas subject to federal regulation. Title 45, Part 46 is "Protection of Human Subjects." Subpart A of this title, referred to as "the Common Rule," regulates the conduct and support of this type of research. All IRB members will need to refer to the definitions and guidance provided in the Common Rule. It requires that, as a condition for receiving federal research support, an institution must establish and delegate to an IRB the authority to review, stipulate changes in, approve or disapprove, and oversee human subjects protections for all research conducted in that institution.[1]

Who Makes Up an Institutional Review Board, and Why?

The composition of an institutional review board depends on what it needs to do an optimal job of review and approval or rejection of research proposals. It must include in its discussions and deliberations as many people as are needed to optimally carry out its work. The membership may include individuals from the institution in which the IRB resides as well as local, regional, national, and even international experts and representatives from patient groups and scientific groups. Calling on experts and representatives is often necessary to clarify both the scientific and the ethical questions and issues that an IRB faces in its daily work.

There are private IRBs, which can be contracted by an institution. However, at present, nearly all IRBs are localized at particular institutions in which research is carried out. This local orientation allows the IRB to capture nuances of issues in research in the particular institution and community.

The *Code of Federal Regulations* specifies many aspects of the makeup of IRBs.

At Least Five Members

The code states that "each IRB shall have at least 5 members, with varying backgrounds to promote complete and adequate review of research activities commonly conducted by the institution."[2] Beyond that, being realistic, the IRB must also be large enough to handle the workload, which will vary depending on the amount and complexity of the research being conducted at the institution.

Sufficiently Qualified, Diverse, and Sensitive to Community Attitudes

The code states that "the IRB shall be sufficiently qualified through the experience and expertise of its members, and the diversity of the members, including consideration of race, gender, and cultural backgrounds and sensitivity to such issues as community attitudes, to promote respect for its advice and counsel in safeguarding the rights and welfare of human subjects."[3] Representatives of the community are included so that they may be informed about the research that is being conducted at the institution and become part of the review and evaluation process, in both its scientific and its ethical aspects.

Scientific Professional Competence, Plus a Lot More

The code states that "in addition to possessing the professional competence necessary to review specific research activities, the IRB shall be able to ascertain the acceptability of proposed research in terms of institutional commitments and regulations, applicable law, and standards of professional conduct and practice. The IRB shall therefore include persons knowledgeable in these areas."[4]

Clinical and Research Pharmacists

Two areas of professional expertise important to have on or available to the IRB are clinical and research pharmacy. Some IRBs combine this expertise in one professional pharmacist; other IRBs may be fortunate enough to have pharmacists with each expertise, one focused on clinical pharmacy, the other focused on pharmaceutical research.

The clinical pharmacist is the link to the hospital's formulary, with all of its prescription medicines and their host of risk factors, side effects, and contraindications. The clinical pharmacist brings an understanding of drug risk, drug-drug interaction in general, and drug–study drug interaction specifically.

Research pharmacists work with new study drugs and are used to working with principal investigators, research teams, and study sponsors in proposal writing and as studies are carried out.

Experts on Vulnerable Participants

The code defines certain groups of study participants as vulnerable: "children, prisoners, pregnant women, . . . handicapped or mentally disabled persons."[5] If the research involves such subjects, the code specifies that the institution consider inclusion of IRB members "who are knowledgeable about and experienced in working with these subjects."[6] According to the code, all vulnerable participants need additional safeguards beyond the protection afforded other types of participants.

In general, additional protection is needed whenever an individual or group may be the subject of coercion. Coercion may be overt or covert and can include subtle forms of manipulation. The IRB must be attentive to even the subtlest forms of coercion, which can range from the information contained in recruitment advertisements or posters to the information exchanged in informed consent sessions between principal investigators and individuals who are interested in participating in or who are being recruited for a study.

Although in the past the code focused on persons who were mentally disabled,[7] more recently the concept has changed to people who have impaired decision-making capacity. Individuals in certain conditions, such as when taking sedatives, in the midst of a stroke, with uncontrolled blood sugar levels, or in the midst of an acute severe exacerbation of schizophrenia, must also be considered vulnerable. Any change in the state of awareness, concentration, or affect to a degree that the person has not previously experienced may be viewed as a change that makes him or her vulnerable in decision making, and this state must be treated as significant and in need of assessment and consideration by the IRB.

Nondiscrimination

The code states that "every nondiscriminatory effort will be made to ensure that no IRB consists of entirely men or entirely of women, including the institution's consideration of qualified persons of both sexes, so long as no selection is made to the IRB on the basis of gender. No IRB may consist entirely of members of one profession."[8]

The Scientist and the Nonscientist

The code states that "each IRB shall include at least one member whose primary concerns are in scientific areas and at least one member whose primary concerns are in nonscientific areas."[9]

The nonscientist(s) on the IRB may have interesting perspectives to share with the scientists involved in the review process. For example, the nonscientist may better appreciate the application of the science to the development of research that ultimately benefits the individual, the community, and society in general or may better appreciate how best to protect vulnerable participants. The nonscientist may reflect on how a participant without scientific training might view a particular study, the terminology used to describe the scientific hypothesis, the scientific methods, and why this study should be done on human beings.

A Nonaffiliated Member

The code states that "each IRB shall include at least one member who is not otherwise affiliated with the institution and who is not part of the immediate family of a person who is affiliated with the institution."[10]

No Conflict of Interest

The code states that "no IRB may have a member participate in the IRB's initial or continuing review of any project in which the member has a conflicting interest, except to provide information requested by the IRB."[11] An IRB member who submits a research proposal should excuse him- or herself from the meeting when the protocol is presented, reviewed, considered, and voted on.

Additional Expertise

The code states that "an IRB may, in its discretion, invite individuals with competence in special areas to assist in the review of issues which require expertise beyond or in addition to that available on the IRB. These individuals may not vote with the IRB."[12]

IRBs and Federal Regulations

With certain exclusions, the *CFR*'s regulations apply "to all research involving human subjects conducted, supported or otherwise subject to regulation by any federal department or agency which takes appropriate administrative action to make the policy applicable to such research. This includes research conducted by federal civilian employees or military personnel, except that each department or agency head may adopt procedural modifications as may be appropriate from an administrative standpoint. It also includes research conducted, supported, or otherwise subject to regulation by the federal government outside the United States."[13]

Within the Department of Veterans Affairs medical centers and research facilities across the United States, the National Committee for Quality Assurance (NCQA) requires that the protection of study participants be the highest priority of each institution and emphasizes the need for each institution to develop a Human Research Protection Program (HRPP) to coordinate the protection of research subjects within the institution. The country is only at the early stages of the development of such programs.

IRBs must not only evaluate all such scientific research when proposed but must review the projects on a continuing basis over time after approval. The frequency of review is at minimum yearly. The IRB will decide for each protocol the frequency of the review, based on issues like the level of risk being borne by study participants. A highly risky study may need to be reviewed many times each year to see how study participants are doing. In addition, a study that an IRB initially marks for semiannual review may require much more frequent reexamination if severe adverse outcomes develop in study participants where those risks were not predicted by the principal investigator, study sponsor, or the IRB's own independent review.

The task of understanding what is required in research on human participants goes beyond the IRB; it is an institutional obligation and applies to IRB members, IRB staff members, the research service, all research staff members, the chief executive officer, the chief of staff, all principal investigators, and all other interested parties. In addition, there must be clear communication with clinical staffs, who must be made aware of the studies in which their patients are participating.

Each member of an IRB should also regard him- or herself as having an expanded role, beyond the dictates of federal regulations. This role is to educate all research investigators within their purview on the best ways to protect study participants with forethought as the investigator conceptualizes a research hypothesis and designs the scientific protocol. This educational job continues as new investigators come to the institution with new staffs, and it needs the dedication of IRB members and staffs.

The Work of an IRB

The basic charge of an IRB is to protect study participants and to judge whether a given research project involving human participants should be allowed to proceed. The process of the board's review can be divided into four categories.

Box 1.1. The Subjects of IRB Review

1. The science and the risks of the scientific protocol
2. The ethical issues in the scientific protocol
3. The informed consent form
4. The informed consent session

The work of an IRB is difficult in part because IRB decision making must be carried out under federal regulations, and often the regulations' general guidance seems insufficient to answer an IRB's questions regarding a particular study. The work is also difficult because it requires the application of ethical standards to research. Furthermore, as I will discuss in later chapters, the ethical principles themselves may need refinement before they can be applied to the specific circumstances of the research.

Lack of Uniformity

The work of the IRB is complicated by the diversity of the circumstances with which it is dealing. There is a lack of uniformity of interpretation of the regulations. There is a lack of uniformity in practices among scientific disciplines; for example, the measurement of the degree of obstruction in the coronary arteries of a study participant with coronary artery disease is a very precise type of measurement compared to the measurement of the degree of aberration in a study participant's mental health. There is a lack of uniformity among scientific protocols in what is being studied and in the project's potential for generating significant new knowledge compared to the risks human participants will be asked to bear in the study. For example, a study of human exposure to lead may be important when viewed in terms of the scientific knowledge to be gained, but have the scientific methods been selected to optimally protect participants and to ensure that any participant's exposure to lead is justifiable. There is a lack of uniformity in subject matter and design employed in questionnaire studies. For example, the risks associated with a questionnaire study of post-traumatic stress disorder (PTSD) may be higher than those of a questionnaire study of whether patients prefer to receive risk information from their health care providers in words (such as "rare, possible, probable," or "definite") or in numbers (such as percentages). This is because the PTSD questionnaire may elicit negative responses from the study participant that could include suicidal ideation or even suicide attempt, whereas the reactions to the preferences questionnaire would likely be benign. There is a lack of uniformity in scientific methodologies; for example, a questionnaire study has a different methodology from a study comparing two drugs or two invasive procedures used in the diagnosis or treatment of disease. There is a lack of uniformity in motivations for involvement in research and for the way a protocol is designed. A study drug might not be compared with the best alternatives on the market because the study sponsor does not want to take the chance of finding that its product does not fare well in the comparison. This lack of uniformity underlying the scientific protocols that IRBs are required to examine and evaluate in accordance with codified regulations is a never-ceasing challenge.

Key Elements Being Reviewed

The task of an IRB can be extremely complex because of the broad scope of the scientific protocols that most are required to review. There are IRBs that review projects within a narrow scope, for example, only in oncology, hematology, and radiation therapy for various types of cancer, lymphoma, leukemias, and other blood disorders. However, most are in academic medical research centers that work on many areas of research, such as cardiology, dermatology, environmental medicine, hematology, infectious diseases, mental health, oncology, public health, pharmacology, radiation therapy, and toxicology. With such diverse academic research departments, the IRB will be faced with protocols and informed consent forms from multiple disciplines.

The process of research begins with a study hypothesis and the inception of a scientific protocol, proceeds to data acquisition and eventual completion of the study, and then moves on to the interpretation of the study data and the generation of new hypotheses and research directions that may be suggested by the interpretation of those data. Throughout this book, I distinguish between the scientific protocol and the informed consent form and session. The scientific protocol contains relevant aspects of the science, including the background research from the peer-reviewed scientific literature, the scientific hypothesis being tested, the time course of the study (including separating clinical care from research), and risks to the participant. The scientific protocol is discussed in depth in Chapter 5.

Box 1.2. Questions Asked during IRB Review of the Scientific Protocol

- What is occurring in the science of the study?
- What scientific question is being asked?
- What scientific knowledge of potential benefit to humans is anticipated?
- What may be the cost of the study in human life or quality of life?
- Is the scientific question being asked important enough to justify asking individuals to sacrifice and accept these risks?
- Is the study designed to best protect human participants?

The informed consent form is a written document specifying key descriptive and quantitative information. The descriptive information includes the terms and concepts related to the research and explains why a study is being done with human participants. The quantitative information includes an estimate of the risk involved. This information is collected in this document to inform any interested party about the nature of the proposed research, the known risks, the fact that other risks that are not known might exist, whether the study sponsor will compensate the participants for hospitalization and/or future disability related to adverse outcomes, and participants' rights with respect to participation in the study, withdrawal from the study, and lawsuit if compensation is not offered. In the vast majority of cases, the informed consent form contains the study rationale and a statement of the scientific hypothesis being tested.

The informed consent session—an in-person, one-on-one meeting between a principal investigator, or someone designated by the investigator, and each individual who is considering participating in the study—is much less understood, as it has not been systematically studied. The informed consent form and session are discussed in detail in Chapter 6.

Box 1.3. IRB Review of the Informed Consent Form and/or Session

Is the scientific study explained in the informed consent form and/or session in such a way that the potential participant understands

- the study question that is being asked, which the individual's participation may help to answer?
- the risks that the individual is being asked to bear and the severity of these risks?
- the likelihood (chance) that the individual may benefit from participating in the study and the nature of the benefit?
- the likelihood (chance) that the individual may not benefit at all from participating in the study and in fact may sustain an adverse outcome?
- the time constraints that the individual is being asked to accept?
- whom the individual should contact at any time with questions about the study (principal investigator or research team member) or about the conduct of the study (IRB chair or a designated IRB member with appropriate qualifications)?
- whom the individual should contact regarding an adverse outcome or injury that occurs during the study?
- whom the individual should contact regarding his or her rights as a participant, such as the right to withdraw from the study?

In examining the scientific protocol and the informed consent form and session, the issues IRB members must address can be grouped into three areas:

- the scientific integrity of the research protocol,
- the level of risk to the study participant, and
- the quality of the informed consent form.

As the members address these three areas, they must assess certain key points of each scientific protocol and informed consent form.

- the purpose and intent of the scientific study being proposed
- the scientific methods and procedures that will be followed
- the risks to humans participating in the research study
- the risks if the level of protection of confidentiality of data is ill chosen or ill conceived and the data are not optimally protected, especially when the data will be uniquely identifiable with a particular individual or his or her family

In assessing each point, the IRB member must ask himself or herself the following questions regarding the study:

- What is the nature of each risk relative to the potential gains of the scientific study?
- Are the risks presented understandably to study participants as well as to the IRB member reviewing the scientific protocol and informed consent form?
- How (using what explicit clinical criteria) will the principal investigator assess the decision-making capacity of individuals being approached to participate in the study? (The best ways to assess decision-making capacity and the adequacy of the clinical criteria are active areas of research.)
- If tools are being used to assess decision-making capacity, how rigorously has the scientific validity of these tools been tested?
- Do individuals who demonstrate initial interest in participating in a research study understand the difference between participating in research and receiving clinical care?
- Whom should the study participant call if he or she has questions about the scientific or ethical conduct of the research study?
- Whom should the study participant call if a problem (adverse outcome) occurs?
- What kind of help will be given the participant for adverse outcomes that occur during or after the study?
- Who will pay for any medical care, medical treatment, or hospitalization for study-related adverse outcomes?
- Will the study participant be compensated for study-related injury and, if so, in accordance with what standard of judgment?

IRB members have to analyze and examine each scientific protocol and informed consent form to make certain they accurately convey the science and risks of the project; the qualifications of the principal investigators, co-investigators, and all relevant research personnel; the terms and disclosures that will be discussed in the informed consent session. IRB members must always be alert in their review, identifying issues that need clarification by principal investigators and formulating questions to ask the principal investigator.

Although the board members start with a submitted scientific protocol and informed consent form, they must go beyond the contents of these to understand what is going on *scientifically* and *ethically* in the study and to determine whether the study abides by federal research regulations. The work of an IRB is complicated by this need to focus on both science and ethics in the decision to approve or reject a study.

The keys to simultaneously attending to both scientific and ethical dimensions are:

- understanding how and when to ask questions of the principal investigator to clarify the scientific and ethical aspects of the research project,
- understanding when to hold discussions with the principal investigator and the research team on how best to protect the human participants involved in the study, and
- understanding the importance of regularly spaced, continuing reviews of the project, which may need to be intensive if a research study is particularly invasive and may carry severe adverse outcomes.

At the completion of its examination and evaluation, the IRB will assign the scientific protocol and its informed consent form one of four verdicts:

- approved,
- approved with minor modifications,
- deferred until key questions are answered, or
- rejected.

Record Keeping

The *Code of Federal Regulations* specifies that the IRB prepare and maintain adequate documentation of its activities.[14] IRB documents include meeting minutes, progress reports, reports of injuries to study participants, and other records related to each approved research study it is monitoring.[15]

The IRB's meeting minutes must be detailed and must include:

- who was present at meetings;
- the vote on actions, including the number of members voting for, against, and abstaining;
- the basis for requiring changes in or disapproving a research study; and
- a written summary of the discussion of controverted issues and their resolution.[16]

In addition to copies of the meeting minutes, the IRB must prepare and maintain copies of all research proposals reviewed, scientific evaluations, if any, that accompany the proposals, the names of the IRB members who reviewed the materials, approved sample consent documents (e.g., informed consent forms), postmeeting documents and communications, all correspondence between the IRB and the principal investigators, and statements of significant new findings.[17] Records are to be maintained for at least three years after the completion of the research.[18]

The Nature of the IRB's Tasks

The IRB's tasks span many levels in relation to society at large, to the medical institution, to the principal investigator, and, most important, to the study participants.

Tasks in Relation to Society

The IRB's primary task of protecting human study participants begins as soon as the IRB receives a scientific protocol and informed consent form from a principal investigator, and often it is not complete until long after the study's end date. Given the unknowns of the research to be undertaken, the task may seem impossible. Indeed, several major institutions in the United States have failed in this task, forcing temporary shutdowns of their research programs. Yes, IRBs can make mistakes, and they need redirection and continuing education, because research on humans is evolving and complex. This redirection and continuing education of IRBs is essential if IRBs are to perform the important work of asking the right questions regarding the scientific hypotheses being proposed, estimating the chance of success of the proposed scientific enterprise, and minimizing conflict of interest regarding the true ends of the scientific study.

Tasks in Relation to the Medical Institution

In the daily functioning of the institution, the IRB may be the only steadfast proponent of optimal protection of research study participants. The best way for an IRB to protect participants is to ensure that everyone in the institution is focusing on the protection of participants, just as everyone in the clinical care part of the institution is focusing on the care of patients. Only through coordinated focus can an institution hope to protect study subjects, because the IRB members cannot be everywhere in the institution at all times. The problem is that whereas the IRB's only focus is the protection of study participants, the institution is also focused on the promotion of research and further development and refinement of the techniques of clinical care. However, successful protection of subjects will also serve to protect the scientists, staff, and the institution and to foster research.

Tasks in Relation to the Principal Investigator

The IRB's charge is to work with the principal investigator to judge whether the proposed research protocol is appropriate and to establish optimal procedures for protecting participants. It does this for each research study that is undertaken in a medical or scientific research institution. While the IRB is providing an independent evaluation of the safety of a study involving human participants, it is also providing insights into ways in which the principal investigator can modify his or her scientific protocols to best protect participants. The IRB expects that the principal investigator will act in response to the assistance and oversight provided by the IRB.

Box 1.4. The Relationship between the IRB and the Principal Investigator

The IRB's tasks are to help the principal investigator understand:

- what is and is not appropriate research on human subjects
- the necessary prerequisites before research may be conducted on humans
- how best to protect each participant during the entire course of the research study

The principal investigator's tasks in relation to the IRB are:

- to understand what the IRB is saying about risks and ways to reduce them
- to clarify any points about which the IRB asks
- to modify the study in the ways recommended by the IRB
- to be willing to withdraw a study that is unacceptable and to seek to understand why the study was considered unacceptable
- to translate the language of the informed consent form into language a potential participant can understand
- once a study is under way, to report to the IRB:
  —any deviation in protocol
  —any adverse event occurring during the study
  —any violation of the protocol

Tasks in Relation to Study Subjects

The IRB's tasks relative to study subjects follow the sequence and design of an approved research study. The first issue is recruitment of individuals to participate in the study. The second issue is the inclusion and exclusion criteria by which participants are selected and the fairness and safety associated with those criteria. (For example, is there a potential for an individual with impaired decision making to enroll in a study without understanding the significance and risks of participation?) One of the last subject-related tasks in which the IRB is involved is deciding whether the individual participant will receive a report of the findings of the study.

The Questions an IRB Must Ask

As they begin to assess each new research study, IRB members should focus on key questions. The answers to these questions will help clarify how the board can direct the principal investigator and study sponsor toward the best protection of study participants.

- How can the IRB protect human participants in this research?
- Is this study important enough to be carried out on human participants?
- What type of study is this?

Let's look more closely at the third question, for answering it will reveal the keys to answering the other two questions.

Types of Medical Research

There are two basic types of medical research. One involves new medical products; the other involves new invasive medical interventions, for example, surgical operations. Medical products research examines the safety and effectiveness of medical devices and prescription medicines. Medical devices include things like heart defibrillators and heart pacemakers. Prescription medicines are drugs that require a physician's prescription in order to obtain them. Research on invasive medical interventions studies procedures that invade or in some way pass through the planes of the human body. They range from heart transplantation to injection of penicillin. If there are two alternative methods for implanting ventricular defibrillators, it may be argued that these two interventions should be rigorously compared in a randomized controlled trial where each study participant gets randomized to one or the other of the implantation methods.

Types of Studies

There are at least four types of studies actively involving human subjects: (a) new study drug versus placebo, an inert substance without intrinsic biologic properties; (b) new study drug versus best alternative prescription medicine now on the market; (c) new intervention versus best intervention now utilized; and (d) intervention 1 now used versus intervention 2 now used.

In determining what type of study is being proposed, IRB members will ask themselves:

- What is the *science* involved (for example, a comparison of patient preferences on an issue, a comparison of two drugs or two invasive interventions)?
- What is the scientific *methodology* involved (for example, a study examining consecutive patients coming to a clinic to complete a questionnaire, a study examining participants randomly assigned to receive one of two drugs or one of two invasive interventions)?
- Most important, what is the severity of the *risk* that is to be borne by the study participants, and what is the likelihood (chance) that any participant will bear a risk of mild, moderate, or high severity? The IRB must specifically address: (1) the likelihood of risk; (2) the severity of the harm that might occur and whether any adverse outcome would be reversible, partially reversible, or irreversible; and (3) the propensity (tendency) of certain individuals or groups to be more sensitive than others and to more readily develop an adverse outcome of a particularly severe nature.
- Can the study can be done with a less-risky methodology? If not, the study's first goal must be to communicate this risk to participants and then to develop the study in such a way that any adverse outcomes are identified, managed, and discussed with the participant and his or her health care provider as quickly as possible.

The Importance of Educating Study Participants

The worth of a research study is judged in terms of the risk borne by humans who may or may not benefit in any direct way from participating in the study. Obviously, then, the education of study participants in the research methods that will be used in the study is an important goal.

Whose responsibility is it to achieve this goal? It is often difficult for a principal investigator to communicate the basics of a research study under consideration, let alone where this particular study fits in the overall range of research study types. The opportunity for education about the spectrum of research should not be missed. Rather, the IRB and research service should consider developing an educational program about the various types of research study being conducted at the institution and the nature of these studies. This educational program would be intended not to relieve principal investigators of the obligation to inform and educate the individuals they are recruiting as study participants but rather to keep the institution focused on its obligation to be honest and open regarding its research enterprise and its goals of developing scientific knowledge to help future generations and optimally protect study participants.

The principal investigator and study sponsor (typically the party supplying the financial backing for the study) must address the three points below not only with the IRB but also with each individual being recruited into the study. In the informed consent form presented to potential participants, the opening statements about the research study should clearly and precisely state the answers to these questions. The opening statements must acquaint the individual considering study participation with three points: (1) the scientific research question being asked in the study, (2) why human participants are needed, and (3) the nature of the study, especially in terms of risks borne by the participants. Individuals considering participation in a study should be free to reject participation based on their reactions to any of these three points, and indeed they should reject participation if they desire, for any reason whatsoever.

Consider this example. One of the most common types of research study an IRB will encounter is a drug study. It might be a study comparing a new drug to a drug used in standard care, comparing a new drug to a placebo, or comparing drug 1 used in standard care to drug 2 used in standard care. Individuals who are considering participating in a drug study will want to know whether they will be receiving treatment (treatment = new drug or standard drug) or whether the study will involve being randomly assigned to receive either a drug or placebo (placebo = nontreatment). Those who participate in a study comparing a new drug to a standard drug or comparing drug 1 used in standard care to drug 2 used in standard care will be receiving treatment, whereas those who participate in a study comparing a new drug to a placebo will have a 50 percent chance of receiving treatment (the new drug) and a 50 percent chance of not receiving treatment (i.e., the placebo). Individuals who are considering participating in a study that involves a placebo must be given information on the controversies surrounding the use of placebo.

Thus, the level of risk to a participant will vary greatly depending on how the study is designed. Any individual considering study participation must understand this point. If an individual considering participation comes out of an informed consent session believing that he or she will definitely be receiving the study drug when in fact there is a 50 percent chance that he or she will be randomly assigned to the placebo group, this is a failure of communication, and the individual's decision regarding participation would have been based on an incorrect assumption by the participant. The informed consent form must be worded in such a way that any individual considering study participation understands that there is a 50 percent chance that he or she will receive the study drug and a 50 percent chance that he or she will not receive treatment (i.e., will receive the placebo). In the informed consent session, the interviewer must ascertain that the potential participant understands this basic notion of chance, of immutable probability. If an individual is not willing to accept a 50 percent chance of nontreatment, he or she should not enter the study. If it is discovered that potential participants do not understand the concept of randomization, then the IRB may want to hold training sessions for individuals considering participation, to explain the basic concepts of research study design, such as randomization. The point is that the individual must be basing his or her decision to participate on correct assessments of the conditions of the study.

The statement of the nature of the study must be open and straightforward, both in discussions between the principal investigator and the IRB and in informed consent forms. Many informed consent forms submitted by principal investigators and study sponsors appear to try to hide the fact that a study involves the use of a placebo. For example, in many informed consent forms, the first occurrence of the term *placebo* is in the methods section. This is a common mistake. Whether a study involves a placebo is a fact that all individuals considering participation need to know right away, because a placebo is nontreatment, and we know that people often enroll in studies hoping to receive the study drug. The key facts of any study involve clear and precise statements of the study's objectives and its methods and what participants will or will not have access to in the way of standard care during participation in the study. This information then needs to be reiterated at various points as they agree to continue participating in the study.

The nature of the study can be clearly stated beginning in the title of the study and can be repeated throughout the informed consent form. The fact that a study involves placebo should be stated in the study title, the introduction, the study methods section, and the alternatives section. Similarly, the fact that a study involves "new drug versus standard drug," or "drug 1 used in standard care versus drug 2 used in standard care" should be stated forthrightly in the study title, in the introduction, in the methods section, and in the alternatives section of the form.

The alternatives section should list the treatments in standard care available to the individual other than participation in any research study and other than participation in a research study involving a placebo.

In summary, individuals considering participation in a research study need to understand the precise nature of the study and why the study is considered to be addressing a question important enough that it is worth placing humans at risk. Potential participants must be accurately and thoroughly informed so that they have a correct understanding of the risks involved in participation and can make a realistic decision regarding their participation.

Weighing the Protection of Study Participants against the Advancement of Medical Science

There is constant tension between two issues inextricably tied to IRB review: the advancement of medical science through studies involving human participants and the protection of the participants. The IRB must weigh the benefits to society against the risks to participants. What is the precise nature of the anticipated benefit to society of a study that requires participants to bear risks? Can this benefit and its worth be communicated clearly to prospective participants?

When a principal investigator comes before the IRB to discuss a proposal for a scientific study, the investigator often defends his or her position by arguing the need to advance science in a particular direction or the need to ease the principal investigator's burden in the protection of human participants. Often, only after the IRB rejects a scientific protocol and informed consent form does the principal investigator become willing to attend to the problems in the study that relate to the inadequate protection of participants.

The IRB will always be faced with this comparison of benefits to society and risks to study participants. One of the key tasks an IRB undertakes is to understand the benefit to society of the studies it reviews. That the study will add to "knowledge," "generalizable knowledge," "generalizable knowledge that may be of benefit to future generations," or some other type of knowledge needs to be argued for, and the principal investigator must logically and scientifically make this argument. The IRB must consider the benefits to society, to the study sponsor, to the principal investigator, and to the institution, and then judge which, if any, of these benefits is worth the risk to human life or quality of life entailed in the design of the study protocol. The new IRB member should recognize that often it will be difficult for the principal investigator or study sponsor to separate the benefit to society in terms of generalizable knowledge from other benefits that may accrue to the involved parties (i.e., themselves). This difficulty on the part of the investigator and sponsor is, by itself, a good argument for the existence of institutional review boards. IRBs spend a lot of time saying, in effect, "Yes, but have you considered . . . ."

Any weighing of participant risk versus societal benefit (or benefit to other involved parties) involves the IRB's recognizing that any research is undertaken at the expense of the risk borne by study participants. This is why it is crucial for the IRB itself to understand and agree that the benefit to society is generalizable knowledge that may be of benefit to future generations. It is not sufficient for the principal investigator or study sponsor to make this claim. The claim must be plausible logically and be scientifically argued pro and con for the IRB to grant the principal investigator the privilege of conducting the research on study participants. That a study participant may be harmed in the course of a study and that an individual may not benefit from participating in a study are key elements that each IRB focuses on as it examines the worth of a research proposal.

In clinical scientific research, because the ultimate goal is to develop generalizable knowledge for the benefit of future patients, the immediate focus is not on the personal care of the individual participant for that person's benefit. For this reason, the principal investigator and the research team must always be watching for any abnormality that may occur in a study participant. The IRB should follow the study closely enough to ensure that the principal investigator and research team are carrying out their obligations to monitor participants closely enough to forestall adverse outcomes when possible and minimize harm to participants if an adverse outcome occurs. During the evaluation phase, the IRB helps identify risks and adjust protocols so as to minimize these risks during each phase of the research. The IRB should reject a protocol that does not optimally protect participants if the principal investigator or study sponsor is not willing to modify the scientific protocol to allow for optimal protection of participants.

When someone who has been a principal investigator becomes an IRB member, he or she may become quite adept at pointing out the protections that are needed in a project in another department but may be less adept at pointing out the same areas of concern in research studies coming from colleagues in his or her own service. Each IRB member must become able to carefully and systematically evaluate scientific protocols and informed consent forms across all services. This procedure is intensive and becomes even more so when a study involves moderate- or high-risk participants or vulnerable participants.

Many aspects of one's point of view can affect one's assessment in weighing benefit and risk of a study. An IRB member must recuse himself or herself from the review and evaluation of any scientific protocol or informed consent form if he or she may have any conflict of interest because of being a colleague, friend, enemy, or subordinate or superior of the principal investigator, study sponsor, or any member of the research team or any member of the sponsor's team. Recusing oneself carries no negative reflection on the IRB member. It improves the objectivity of the board's decisions about that particular proposal.

The struggle between science and ethics will always exist in the IRB because the IRB's primary focus is on the best protection of humans participating in scientific research. To begin developing an approach to best protect study participants, the IRB member must understand what the research study is about and why the principal investigator and study sponsor believe that this scientific research question should be studied in humans. This struggle between whether the research is worth conducting in humans is always at the heart of any decision that an IRB must make. It becomes especially important on close votes and on those decisions that seem to divide the IRB evenly over rejecting or approving the study proposal. The issue of close votes on IRBs deserves concentrated scrutiny and would make an excellent subject of research on IRBs.

2 Basic Terms and Concepts Used in IRB Work

The basic terms and concepts in IRB work are distinctive in many ways.[1] Even experienced institutional review board members may struggle with what a specific regulatory term means as it is to be applied to IRB work. In 1974, the National Commission for the Protection of Human Subjects of Biomedical and Behavioral Research was created to identify the basic ethical principles that should underlie the conduct of research on humans in the United States. The Belmont Report, published in 1978, summarizes the commission's conclusions. That report and the *Code of Federal Regulations* provide a framework of definitions and guidelines for disclosing information in research and for protecting study participants.[2]

However, the notion of what constitutes scientific research is evolving, and more scientific activities are being brought before IRBs. Much of the current guidance offered to scientists by regulators is to take the question to the appropriate IRB for a decision. Consequently, IRBs are in a delicate position, because they have to grapple with concepts and terms that even the writers of the Belmont Report found difficult. This is not to say that IRBs must undertake a wide-open redefining of concepts, but an IRB needs to clarify for itself many of the concepts and terms that appear in the Belmont Report and the *Code of Federal Regulations*, as they apply to the questions that come before the IRB. Although regulatory agencies may offer guidance, IRB members must share an understanding of these concepts and terms in order to make their decisions.

Basic Terms in IRB Work

IRB work is filled with basic terms. I will review the basic terms individually and in combination with related terms used in understanding the science and the ethics of research involving human subjects. Although these terms are basic, they refer to concepts that in many ways are problematic; these difficulties will be addressed in the second half of this chapter.

The IRB must make certain that all terms are defined explicitly at the outset of any conversation or communication so that there will never be confusion among the parties involved.

Basic

The word *basic* has many meanings, such as "of primary importance" (e.g., "basic scientific truths"). *Basic* can also mean "serving as a starting point" (e.g., "basic education" or "basic tools or methods for conducting scientific research"). Or *basic* can refer to an introduction to an approach or a discipline (e.g., "a basic course on scientific methodology"). *Basic* can also pertain to a foundation, that is, something "fundamental" (e.g., "a basic scientific fact" or "the basic constituents of a larger complex object, form, or living organism"). Finally, *basic* can indicate that something is reduced to the simplest and most significant form, the essence (e.g., "a basic scientific truth or canon").

Science

There are many types of science, but no matter what type, the essence of science is measurement. In observational sciences, what is being watched and measured may be the behaviors of individuals, either alone or in groups. In interventional sciences, what is first being watched and measured are baseline states, that is, states before intervention. After an intervention, additional measurements are taken to assess what changed (positive and negative) due to or in association with the intervention.

The behavioral sciences focus on human actions. The behavioral sciences include, but are not limited to, anthropology, psychology, and sociology.

The medical sciences and the behavioral sciences—through diagnosis and monitoring of disease before and after treatment—try to understand how and why diseases occur and progress and what therapies can slow, prevent, or cure disease processes. Research in these sciences strives to, for example, develop new medical screening, diagnostic, management, and treatment techniques or interventions and compares them with standard medical techniques or with nonmedical interventions or styles of management.

In one area of medical science, genetics, the science and ethics of research have yet to be clearly demarcated. Genetic studies involve not only active research but also the storage of human material—cells, tissues, and organs—for future research. In genetic research, the IRB is concerned with not only the protection of the immediate participants (e.g., in a clinical trial) but also the protection of present *and future* generations.

Scientific Methods

Scientific methods are the principles, processes, and techniques that underlie formal inquiries into the external world in general and, in medicine, into the biochemical, electrical, biologic, physiologic, biomechanical, and other phenomena that make up human physical and mental life. At minimum, the cornerstones of scientific investigations include the tasks of hypothesis formulation, observation, experimentation, and measurement. These are then followed by data recording, analysis and interpretation of the data, and, finally, the deriving and reporting of conclusions from the data. These conclusions are then analyzed and critiqued by other scientists.

Hypothesis formulation should be clear and straightforward but can be confounded by conflict of interest, influencing what the study will examine with respect to the medical product or intervention. Observation of the study subject starts with examination of the subject's baseline condition. The subject may then undergo an intervention or receive a newly developed drug, then more observations are made. Many times, it will be hard to separate the processes of observation and measurement because the observations may be measurements that may be either direct (e.g., using devices to monitor blood pressure) or indirect (e.g., asking how the subject feels after ingesting the drug). Observation and measurement may be limited by what can be ethically and morally sanctioned as a suitable intervention for a person to undergo as a study subject. Certain measurements cannot be obtained fully in human beings because of the length of time and duress the study subject may be placed under. Therefore, lesser measurements are considered to be acceptable because of the morally objectional nature of more invasive interventions.

Measurements and observations can be confounded by not including enough of them at the right times to allow a clear distinguishing of results. These confoundings continue through the scientific review processes and as the editors and reviewers of peer-reviewed medical journals consider if the scientific methods used to test the hypothesis were so selected that valid conclusions can be drawn from the data. In the end, each scientific study is just a step in our understanding of the generalizable knowledge that can be applied to the care of humans.

Intervention

A *diagnostic* intervention is used to identify the etiology of a person's physical signs or symptoms. A *screening* intervention is used to identify early disease in someone who does not have symptoms related to that disease process. A *treatment* intervention is used to cure disease in a patient or improve his or her health.

An intervention may be *noninvasive* or *invasive.* With noninvasive interventions, there is no direct penetration of the body. Examples include ultrasound imaging (in which an organ or other part of the body is studied with the use of sound waves), taking blood pressure with an externally applied cuff, and listening to heart sounds with a stethoscope. An invasive intervention is a procedure that penetrates the body. Many invasive interventions used today in routine clinical care have been developed and refined through research involving humans. Examples include upper endoscopy, colonoscopy, cystoscopy, and catheterization to measure blood flow (for example, cardiac catheterization to visualize the coronary arteries involves the injection of a dye).

Invasive interventions always carry more risk to the participant than noninvasive interventions, and individuals need to understand this risk when they consider participating in a research study involving an invasive intervention (as they should when undergoing such procedures in the course of clinical care).

Research

Research is difficult to define. Approached in terms of its goal, research is "the intended development of new scientific knowledge for future populations." It is "intended" because there is no guarantee that the research will yield new scientific knowledge. It is "knowledge for future populations" because the individuals participating in the study may not immediately benefit, although they may. A case in point is the participant randomly assigned to receive a placebo (an inert substance without intrinsic biological activity): this individual will not benefit from the new knowledge gained from the study because that knowledge will depend on the analysis and interpretation of the data from the study, and that process likely will occur after the individual's participation is over.

Unfortunately, study participants may have difficulty sorting out the different terms that can be used synonymously with *research* (for example, *research study, clinical trial*). Nonscientists usually see more distinctions among these terms than do scientists, researchers, and IRB members. The equivalency of terms needs to be specified when nonscientists and nonresearchers draw distinctions that are not intended by the scientists, researchers, and IRB members using the terms.

Research, as defined by the *Code of Federal Regulations,* is a "systematic investigation, including . . . development, testing, and evaluation, designed to develop or contribute to generalizable knowledge."[3] The code contains the important statement that "development, testing, and evaluation" of tools are also key components of research.[4] Research is traditionally contrasted with two other activities: clinical care and innovative therapies.

Clinical Research

Clinical research is the study of human volunteers in relation to health, disease, and the diagnosis, cure, management, prevention, and treatment of disease, with the goal of developing generalizable medical and scientific knowledge that may be of use for the health and care of future generations.[5] Additional elements of clinical research will be noted in a later section, where it is distinguished from clinical care.

Basic Research

In contemporary science, the term *basic research* has several meanings. One accepted meaning is "scientifically fundamental research," that is, research into the fundamentals of science, such as biochemical or molecular genetic studies, and in this use it refers to laboratory or "bench research," research that is carried out in the laboratory instead of on the physical and mental states of living human organisms. Basic research can also be found in the computer sciences, mathematics, and the social sciences, among other areas.

This book focuses on basic research. Any field, especially a field that relies on a precise methodology, will use the word *basic* to connote a fundamental level of approach to and understanding of a problem. Fundamental levels of understanding can change with the development of a new tool, such as a visualization device that offers a way to measure something not previously measurable. For example, early research using the PET scan has allowed investigators to measure what parts of the brain are involved when a person is lying as compared to telling the truth.

When I work with new IRB members, I use the following definitions of *basic research.* First, basic research can employ observation only (*observational research*) or observation and intervention followed by further observation (*interventional research*). Second, basic research can be hypothesis *generating* (that is, looking at data without having a scientific hypothesis but developing a hypothesis from the data, sometimes referred to as "going on a fishing expedition" or "data fishing") or research can be hypothesis *driven* (starting with a scientific hypothesis and testing it against the data that is gathered in the study).

Box 2.1. Observational versus Interventional Research

Basic observational research involves the following sequence of activities:

1. observing
2. recording data (at a time and over time) related to the observations

3. interpreting results
4. drawing conclusions, which then are debated in the peer-reviewed medical literature, generating modifications to the hypothesis of the study and further research
5. verifying and validating conclusions

The conclusions, once verified and validated, can lead to the development of interventional studies in clinical research involving human participants.

Basic interventional research depends on observation but goes beyond observational studies by intervening after an initial observation period. Interventional studies involve the following sequence of activities:

1. observing before the intervention is undertaken
2. recording preintervention data
3. intervening
4. recording intervention and postintervention data
5. interpreting results before and after the intervention then continuing the sequence of activities as noted above

Recruitment

Recruitment refers to inducing an individual to consider enrolling in a particular research study. Recruitment may start with an IRB-approved advertisement that the study is seeking participants. This IRB-approved advertisement may appear in a particular area of a hospital or medical center that has been approved by the institution as a place where individuals wanting to consider what studies are open for participation may see what is available in this particular institution. IRB-approved advertisements may also appear in newspapers, on the hospital or medical center's websites, and other locations. These are neutral domains of recruitment, where persons who are interested in research may seek out further information.

Recruitment takes on a different sense when a health care provider offers his or her patient information about a research study for which he or she is the principal investigator. Here the clinician-investigator is placing the potential study recruit at a disadvantage. Many patients want to discuss with their clinician whether or not to enroll in a research trial. If the patient's clinician is conflicted in the sense of being involved in the study in some manner, that clinician cannot offer the independent opinion that the patient seeks. In the case of such recruits, the IRB must be concerned not only with real conflict of interest but also the appearance of conflict of interest.

Benefit

A benefit is a gain of some sort. In clinical research, an individual may not directly benefit from and may in fact be harmed because of study participation. This point must be made clear to anyone considering participation in research. The benefit patients are seeking in clinical *care* is an improvement in their state of health or a reduction in their risk of developing disease. Benefits of participation in clinical research are often difficult for the individual considering study participation to understand because the purpose of research is the development of generalizable knowledge that may help *future* generations. Some research volunteers feel that they benefit from the research *because* they are able to help future generations, but these individuals must also recognize the risk they bear as study participants. Clinical research is not clinical care, and study participants must understand the distinction. Clinical care and clinical research are further distinguished below.

Risk

Risk is a word that is used in many senses. Sometimes *risk* means uncertainty as to outcome or result. Often, *risk* refers to the degree of uncertainty involved in an outcome. Sometimes *risk* means the chance or probability that an adverse outcome will occur, and sometimes it means the adverse outcome itself. Practically, in clinical research on humans, there is risk of harm, risk of not being treated, risk of receiving a placebo, risk of health not being improved, risk of loss of privacy, risk of nonmaintenance of confidentiality or anonymity, and more types of risk.

In IRB work, *risk* is used in at least five senses in relation to adverse outcomes: (1) the nature of the adverse outcome, (2) the severity of that adverse outcome, (3) the chance or probability that an adverse outcome will occur, (4) what can be done to minimize the chance that an adverse outcome will occur, and (5) what can be looked for to detect as early as possible an adverse outcome or any of its preconditions so that action can be taken to avert or minimize the harm or bad result in a study participant.

Risk of harm to the study participant is the dominant risk that an IRB will consider in its evaluation of a research proposal. The goal is the minimization of risk to the participant and the early detection and management of those risks that are considered acceptable in the study. Yet, many different arguments can be made regarding acceptable risk, and a risk that is considered acceptable by one person may be considered unacceptable by another person.

The notion of risk of harm to study participants includes both physical and mental harm, risks related to data and loss of data that can then be tracked back to participants, and risks of loss of employability, loss of insurability, and social stigma that may occur with disclosure of genetic data. Discovery of genetic data can lead to problems not only for the study participant but also for blood relatives in past, present, and future generations.

Minimal Risk

The notion of *minimal risk* should be interpreted as meaning minimal for that participant in a safe environment. The problem with this interpretation is that persons have different capabilities even within the same safe environment. For example, in a safe environment, daily activities such as walking across the street can be considered as carrying minimal risk. However, for participants who are physically challenged, even in a safe environment, walking across the street might carry more than minimal risk. Risk cannot be interpreted as minimal or acceptable without understanding the practical definition for the particular participants and what they *consider* as minimal risk in their own lives.

The IRB must be alert to what a principal investigator or study sponsor is calling minimal risk. In many senses, the word *minimal* is too subjective to address the real risks involved in a research circumstance or setting. Degree of risk should be considered with the perceptions of multiple study participants from the perspectives of what the risks are, what *risk* means to them, and what outcomes they are willing to accept in the short, medium, and long terms.

An example of a minimal-risk study is a medical chart review in which there is no identification of a patient by a unique identifier, such as a name, Social Security number, or hospital record number. However, if there is a unique procedure number, such as a unique number for a colonoscopy, this leaves open the possibility of identifying a patient. A study in which colonoscopy was performed on participants would be considered a medium- or high-risk study, depending on the participant's medical history and current health circumstances.

Another example of a minimal-risk study is a short questionnaire in which there are no unique identifiers and the questions do not concern illicit behaviors or sensitive subjects (domestic violence, for instance).

Clinical Care

Clinical care is the medical and health care that patients receive. The goal of clinical care is to maintain and/or increase patients' health, chances of survival, and quality of life over time in regard to preventable, diagnosable, and manageable or treatable medical conditions or diseases affecting their body and mind or affecting the quality of their lives. Clinical care and clinical research share many elements. For example, the clinician who sees a patient may also be a principal investigator on a clinical research study. Clinical care providers may also be co-investigators or collaborators in a research study. The clinical setting where patients come for medical care may be the same room where individuals are recruited to participate in a research study.

In such a setting, where physical examinations are carried out and blood is drawn for both clinical care and research studies, it is not surprising that individuals may become confused about what is care and what is research. It may be difficult for the individual to distinguish when his or her physician or nurse is acting as care provider and when as research study investigator. Without specific, clearly stated reminders, an individual may become confused as to why he or she is in a clinic—as a patient or as a study participant.

There is risk and uncertainty in decision making in both clinical care and clinical research. A key aim of clinical care is to benefit the individual medically, but this is not the aim of clinical research. Again, as stated in the *Code of Federal Regulations,* research is "a systematic investigation . . . designed to develop or contribute to generalizable knowledge."[6] Research is aimed at benefiting future populations, not the population being studied.

Innovative Therapies

Therapies are interventions aimed at treating and managing diseases or medical conditions. Innovative therapies, often called "experimental therapies," are newly devised ones. They are not systematic investigations, which means they are not technically research. Research involves groups of carefully selected study participants, but therapies are used with individual patients. Sometimes, for instance if there are no other approaches to try to help a patient's condition, an innovative or experimental procedure will be used therapeutically, to try to make the patient better. Although federal regulations permit this use of innovative therapies, I believe that, for the IRB's purposes, innovative therapies should be considered clinical research. The development and design of innovative procedures and techniques should be systematic and should afford the patient the same protections afforded a study participant.

Coercion

Coercion is the placing of pressure on an individual to do something for reasons other than his or her own. Coercion can be subtle: persuasion, argument, and personality can be used to compel an individual to act in a certain way. A health care provider's attempt to influence a patient's decision to participate in the provider's research study could in some circumstances be considered coercive. Coercion—including all the subtle forms—has no place in research in general and in research on humans in particular.

IRB members must recognize coercive language when they see it. One of the main tasks of the IRB in reviewing an informed consent form is to make certain that any coercive language is eliminated. The informed consent form should, for one thing, contain no language even remotely suggesting that the participant is giving up the right to bring suit against a principal investigator or study sponsor for an adverse outcome that occurs during participation in research. The description of alternatives in an informed consent form should contain no phrasing or content that makes it appear that few or no alternatives are available in standard medical care if that is the case. Any phrasing that reduces participants' rights is not allowable. Whenever it appears that the principal investigator is manipulating an individual's decision toward study participation through the use of language that seems to limit the consideration of available options, coercion is present, whether it was intended or not. The way in which benefits are discussed can also be coercive, inducing an individual into a study by implying that there will be a benefit to the participant, when in fact there may be no benefit to him or her, or by exaggerating the potential benefits for future populations.

Data

It is the IRB's job to ensure that the principal investigator is always vigilant about protecting the privacy of data that are labeled with unique identifiers. The IRB can routinely check principal investigators' offices to verify that data are secure while being viewed, analyzed, and interpreted and are stored in the most secure fashion possible both during and after the study. It is also the IRB's job to ensure that study participants are informed of the risks that could occur with the release of any sensitive data gathered in the study and the fact that, no matter how secure the data may appear to be, there is some chance of such a release.

Linked data are data that can be connected to an individual indirectly, through a record or codebook. *Unlinked data* are data that can never be traced to an individual by any means. *Identified data* are data that can be directly connected to an individual, without any decoding, and exist as data derived from this specific individual. *Deidentified data* are data that at one time were linked to an individual but later were stripped of the identifying linkage. Deidentified data may or may not be traceable back to the original data set from which they were derived.

Privacy, Anonymity, and Confidentiality

In clinical research as in clinical care, a participant retains the rights of privacy, anonymity, and confidentiality. In both arenas, even if the best attempts are made to keep data confidential, there is still a chance—although small—that those data will become public (for example, through theft of data or data files kept on paper or electronically).

The safest way to protect participants' privacy is to make all data anonymous. *Anonymous data* are linked to no unique identifiers. *Unique identifiers* are labels used to identify data that contain information that can link it to a particular individual. Unique identifiers include such information as an individual's name, Social Security number, or hospital record number. *Non-unique identifiers* provide no unique linking between the data and an individual. Examples of non-unique identifiers are random study numbers. However, if a random study number links a participant to a unique identifier in a codebook, this codebook must be treated with the same security as a unique identifier. If the codebook is stolen, then the participant is placed in jeopardy regarding the use and potential misuse of that information.

That basic terms are ill defined or poorly illustrated in research proposals and that IRBs must make judgments for the protection of participants without clear and precise workable definitions are reflections of the rapidity with which human research is changing in the United States today. These basic definitions must be the subject of continual IRB study. The regulators who review IRB decision making will ask the IRB to define its understanding of the concepts and terms it is using and to specify how it is viewing, interpreting, and defining concepts in its primary work, protecting study participants.

Basic Concerns of IRB Work

Conflict of Interest in an IRB Member

Conflict of interest in general is an improper balancing of interests. Conflict of interest on an IRB most typically happens when a member (or set of members) of the IRB is improperly weighting private interests against the position of trust to protect human subjects of research. An appearance of a conflict of interest must be considered a conflict of interest, and there can be a conflict even when the individual does not recognize it as such.

The presence or even appearance of a conflict of interest on an IRB may cast doubt on the fairness of the board's decision and open an IRB to attack on the grounds that it made a decision for reasons other than the best interests of the study participants. If an IRB is charged with a conflict of interest, it should examine its ability to function and perform duties.

An individual IRB member may not always recognize that he or she has an apparent conflict of interest. The whole IRB must be attentive to such appearances and bring them up for discussion. The IRB must recognize a potential conflict of interest, declare it openly, and then eliminate the conflict or appearance of conflict by excluding the affected IRB member from deliberations on that study.

Sometimes a conflict of interest occurs when an IRB member has information but does not share it with the IRB, for example, that there is an existing or anticipated relationship between the member and the principal investigator, research team, study sponsor, or other party. The relationship interferes with the member's making an unbiased assessment of the study under consideration. A member might be protecting his or her employment in the institution by not disclosing facts that would shed a negative light on the scientific proto-

Box 2.2. Conflict of Interest in an IRB Member

Each IRB member should ask him- or herself, Am I involved in any way—financial, professional, or personal—with the sponsor, principal investigator, or anyone on the research team, and could that compromise my decision making on the IRB? This may be a real conflict or just have the possibility of appearing to be a conflict to an outside observer.

Examples for consideration include:

- owning stock in the field in which the research is being conducted
- owning stock in the company that is sponsoring the research study or in one of its subsidiaries
- a relationship with the principal investigator, co-investigator, collaborator, or member of the research team
- a personal relationship with anyone involved in conduct or sponsorship of the study that could compromise the member's decision making on the research study
- outcomes of the study that would affect the member's career-related advancement

col or informed consent form being reviewed. Say a principal investigator failed to read the study sponsor's scientific protocol and informed consent form and submitted them to the IRB as received. The IRB member who was aware of this and failed to tell the IRB would be withholding information that not only might help the IRB understand what the principal investigator was doing but also might help the principal investigator recognize and change behaviors that could result in harm to a study participant because of failure to understand the details of the scientific protocol.

A financial conflict of interest could involve, for example, an IRB member's owning stock in a pharmaceutical company that is sponsoring a study the IRB is considering. Institutions may differ on the number of shares that can be held by an IRB member before they consider the interest problematic.

Conflict of interest may arise in the relationship between a worker and a supervisor, between co-workers, between colleagues, or between enemies. A conflict of interest may be based on past, present, or anticipated relationships or dealings between any of these parties. For example, an internist sitting on an IRB may be in a conflict of interest when the chair of his or her department proposes a study; a nurse sitting on an IRB may be in a conflict of interest when the director of nursing proposes a study.

If a principal investigator is a colleague of an IRB member there may be conflict of interest or the appearance of same. Obviously, if the IRB member is a co-investigator or collaborator with the principal investigator on another study, the conflict would be even more likely.

If an IRB member is in a position of real or apparent conflict of interest regarding a particular study, the member should declare that conflict of interest to the IRB and offer to recuse him- or herself from the discussion and vote on the study. The board will discuss the issue. A recused member may be called back to answer questions or provide further clarifying information but then must leave during the subsequent discussion and vote. If an IRB member remains in the room while discussion is taking place (for example, when the IRB member's own research proposal is being considered), the discussion is bound to be constrained by his or her presence. This is a problem, because it means that IRB members are holding back. Discussion is key to helping the IRB make a decision in the best interests of the participants. The recused member will be informed of the IRB's decision in regard to the study. The minutes of the meeting should note that an IRB member with a declared conflict of interest regarding a particular study recused him- or herself during the discussion and vote.

If an IRB member thinks he or she may have a conflict of interest but isn't sure, that member must raise the issue with the board for consideration. The

IRB may need to consult local, regional, or national experts in the area of conflict of interest, just as it does in areas of scientific or ethical inquiry. When a member must recuse him- or herself due to conflict of interest, it may be necessary to call in another similarly qualified person to replace the recused member, so as not to leave the board without coverage in that area of expertise or experience.

Participant Safety

The safety of and prevention of harm to study participants is the primary concern of IRBs. Participant safety has five main components:

1. elucidating what adverse events might happen to a participant
2. preventing adverse events from happening to a participant
3. monitoring laboratory and study results so as to detect abnormalities as quickly as possible
4. communicating with participants about laboratory and study results, so that study participants know as soon as possible when an abnormality is detected and needs to be acted on
5. managing abnormalities as quickly as possible for the best clinical care of the participant

The IRB's concern for participant safety starts when they review the scientific protocol and informed consent form. The review should include what is occurring in informed consent sessions between the principal investigator and potential participants. In addition, the IRB must focus on participant safety through the systematic review and exploration of any complaints participants may file with the IRB.

No delay is allowable in the systematic review and interpretation of laboratory and study data and action on any abnormality discovered. If an abnormality creating a potential for harm to the participant is identified in a test, study, or questionnaire, the result must be acted on for the optimal care of the participant. For example, the identification of depression in a participant through a quality-of-life questionnaire used in a research study must be acted on: the research team must address the participant's depression, discuss it with the subject, and refer him or her to an appropriate health care provider. The management of this participant within the study, given a potential new diagnosis, must be discussed with his or her primary care provider.

Participant safety is accomplished in part through the performing of laboratory and study testing, but it also may require that qualified experts review the results and communicate with both the participant and his or her clinical care physicians to ensure follow-up, management, and continued care. Any new abnormal laboratory test or study result must be addressed, both in the research setting and in clinical care. In many cases, the abnormal result will need immediate action. An investigator must be certain that the research team has clinicians associated with it who are trained to act immediately on any adverse outcome and to communicate the fact of the adverse outcome to the participant's primary care or other physician. The goal is to identify any adverse outcome as early and as quickly as possible and to act in such fashion as to minimize any harm to the participant. For example, a new drug being tested in a clinical trial may cause problems with the functioning of the kidney and/or liver (the major organs of metabolism of drugs in the human body). The research team must communicate to the participant, in clear language, about any abnormalities in kidney or liver function as soon as they appear, explaining how the abnormalities developed. The team must notify the participant's health care provider.

Concern for participant safety continues throughout the study. In a multisite study, information from the study sponsor (often a product manufacturer) is reported to the IRB from other sites on a regular basis. The IRB must decide whether and how to communicate new information about risks and abnormal occurrences at other medical centers to the participants enrolled at their own medical center. The IRB continues reviews of a study until the time of study closure. The process may continue beyond that point if new results about benefits or unanticipated risks are revealed, and these may need to be communicated to participants.

The final area of concern over participant safety involves unlikely but possible severe risks. The IRB must ensure that the principal investigator and study sponsor provide to the individual considering participation an informed consent form that contains a full and lucid disclosure of all severe adverse outcomes with the best estimates (from the peer-reviewed medical literature and from expert clinical and nonclinical opinion) of the chance or likelihood of occurrence to the individual study participant. The IRB can ensure that a full disclosure takes place in the informed consent form, and they need to emphasize to the principal investigator and the person who will hold the informed consent session that the study risks not be minimized.

Informed Consent

Informed consent is usually considered to be a process, occurring over time. After an individual is made aware of the existence of a research study in which he or she would be eligible to participate and the individual expresses interest in participating, there is an informed consent session between the individual and the principal investigator or the principal investigator's designee. Prior to the session, the potential participant has been given the informed consent form to read carefully. At the end of the session, the individual is asked if he or she would be willing to participate in the study based on what was said in the informed consent session, what is written in the informed consent form, and the individual's understanding of the information communicated.

The *Code of Federal Regulations* allows one or more members of the IRB to observe informed consent sessions, but at present this step is seldom taken. For new investigators conducting an informed consent session for the first time, observation by an experienced interviewer, followed by a critique, can provide helpful training. Even experienced investigators may benefit from observation and criticism aimed at helping them best communicate with individuals regarding the nature, risks, alternatives, and details of study participation.

Materials Reviewed

The main materials the IRB has to review and judge are shown in Box 2.3. The content of these documents will vary greatly, depending on the nature of the study. For example, a study proposal involving the storage of tissue for use in future genetic research must clearly spell out the requirements for informed consent for genetic research that will be consistent with both federal and state laws.

After initial review of the study, the IRB can generate a set of questions to ask the principal investigator, the study sponsor, or others, to complete its understanding of the science and the ethics of the study. The principal investigator usually serves as the primary communicator with the IRB, the study sponsor, and others. The IRB can bring in as many experts as necessary to better understand the science and ethics of a study.

Box 2.3. Basic Materials Reviewed by IRB

For initial review of a study

- Research project application
- IRB application or initial review questionnaire
- Scientific protocol
- Informed consent form
- Recruitment brochure, advertisements, etc.

- For drug studies, investigational new drug (IND) numbers, etc.
- For device studies, investigational device exemption (IDE) numbers when appropriate
- Any reports to the IRB in regard to the study and its investigators
- Any prior decisions by the IRB in regard to the study and its investigators

For review after approval of the study

- Reports of adverse outcomes, on site or off site
- Any amendments, changes, or modifications to the study
- Continuing review materials
- Any participant's complaints

Problematic Concepts

What Is Research?

The question that must be in every IRB member's mind is whether the activity he or she is reviewing is in fact research. There are various working definitions of *research.* The Belmont Report gives the following description: "the term 'research' designates an activity designed to test an hypothesis, permit conclusions to be drawn, and thereby to develop or contribute to generalizable knowledge (expressed, for example, in theories, principles, and statements of relationships). Research is usually described in a formal protocol that sets forth an objective and a set of procedures designed to reach that objective."[7]

The Belmont Report refers to hypothesis-driven research, that is, "an activity designed to test an hypothesis."[8] Do other types of scientific activity, which might be characterized as "non-hypothesis-driven" research, fall under the purview of IRB review and evaluation? An example of non-hypothesis-driven research might be what is loosely termed "data examination" or "data fishing," that is, taking a data set that was formulated without a specific hypothesis and then examining relationships among its variables for the purpose of generating a hypothesis. Reasonable people differ on whether or not this type of examination, analysis, and interpretation of data should be considered research and whether it should be subject to IRB approval.

If research is conducted on such a data set, particularly if the data is about patients, the IRB must ensure that patients' permissions have been obtained properly and substantively. In addition, there must be a clear statement of who has access to the uniquely identified data set and for what purposes. The patients must be made aware of all parties who would ever have access to the data set and of all purposes for which the data set would be used. Some people argue that when data are not uniquely identified (that is, when there are no links to a unique individual through name or unique number) there are fewer potential problems with the use of such data and less need for stringency in standards of its use.

Clearly, it may be difficult to determine, at the time a physician is initially accumulating data, whether those data will constitute research, especially if a hypothesis is not present. Questions about the physician's intent can be raised at each of the many stages along the data pathway.

1. What was the intent of the initial data collection?
2. When did the physician first have the idea that the data set could be examined to look for relationships among variables in a hypothesis-generating type of activity?
3. How will the IRB assess the physician's intent?
4. How will the IRB view this hypothesis-generating type of activity within its overall understanding of research and IRB decision making?

Additional questions arise when the IRB is asked to consider a case series. A case series is a collection of data about patients seen, for example, in medical clinics, in departments of medicine, in surgery clinics, or in departments of surgery, where "associations" among data are being made for interpretation and discussion. For example, a case series may include the first 50 or 100 patients at a particular medical center who have a specific infectious disease and were given an antibiotic recently approved by the FDA. The case series may identify the characteristics of the patients in relation to a set of parameters and may serve as a starting point for further research on use of this antibiotic with a much wider range of patients. The questions now arise: Is this case series to be considered research or only the prelude to research? If the latter, should it fall under the purview of an IRB as research activity even though it is not necessarily testing a hypothesis?

The Belmont Report acknowledges that problems exist in the basic definition of research—what is contained inside the boundaries of research, what is excluded as not research, and what scientific activities sit on the boundary awaiting placement on one side or the other. The report begins to approach the issue of how the IRB balances competing perspectives (pro and con) in its determination of whether or not any particular scientific protocol should be considered scientifically worthwhile.[9] I believe that as many scientific activities as possible should come under the review of IRBs. Yet, IRBs do not take a uniform approach to answering the preceding questions.

The Belmont Report further characterizes research by distinguishing between research and "innovation in standard or accepted practice." "When a clinician departs in a significant way from standard or accepted practice, the innovation does not, in and of itself, constitute research. The fact that a procedure is 'experimental' in the sense of new, untested or different, does not automatically place it in the category of research."[10] However, the report also recognizes that "radically new procedures of this description should, however, be made the object of formal research at an early stage, in order to determine whether they are safe and effective."[11]

The notion that a procedure that is experimental (in the sense of being new, untested, or different) is not automatically considered research is complex; yet, it is a fact of everyday life in the care of patients, particularly in urgent or emergent circumstances. Take the case of a patient who has a large abdominal aortic aneurysm that may burst at any time, causing death. This patient is evaluated by surgeons and an interventional radiologist, who find that the patient's long history of cigarette smoking has caused such extensive damage to the lungs that the risk of anesthesia is so great that he is not considered a candidate for surgery. The interventional radiologist believes he can insert a device that may allow the patient to live. During the procedure to insert this device, the interventional radiologist finds that the marketed device will not fit the patient's unique anatomy. The radiologist, on the spot, creates a new device (or significantly modifies the existing device) to fit the necessary dimensions and then, under sterile conditions and after sterilizing the device, inserts the device in the patient. The patient survives. This circumstance could not be categorized as research, but it can lead to further research on ways to generalize the knowledge of how to modify devices. Thus, this experimental device, born sheerly out of necessity, may be the first step in the development of a research proposal that will aim to develop ways to modify standard devices to allow their safe and effective use in patients with nonstandard anatomy.

According to the Belmont Report, "research and practice may be carried on together, when research is designed to evaluate the safety and efficacy of a therapy. This need not cause any confusion regarding whether or not the activity requires review; the general rule is that if there is any element of research in an activity, that activity should undergo review for the protection of human subjects."[12] The report failed to attend to the possible confusion that can be generated in a patient's mind when, for example, the clinical care setting in which the patient sees a physician is also the research setting where the physician is the principal investigator of a research study in which the patient is enrolled.

Additional questions left unanswered by the Belmont Report are: How can an IRB be sure whether or not there is an element of research in an activity? Does the fact that the data will be or have been systematically collected constitute an element of research? (Many of a medical institution's quality-assurance activities involve the systematic collection of data and are not considered research, because they do not necessarily lead to generalizable knowledge.) Although I believe it is best to bring all such activities to the IRB for deliberation and to consider them research, this belief is not uniformly held (for example, by busy medical research institutions in which the IRB's time and energy are already stretched to the limit by the systematic study, review, and evaluation of hypothesis-driven research). These questions about non-hypothesis-driven research need further thought.

Weighing the Value of Study Results against Risk to Human Subjects

The IRB's main work, the protection of human study subjects, is difficult, in part because of the tremendous variation in the interpretations of which scientific hypotheses are important enough to pursue by experimentation and research in humans. The IRB member should anticipate receiving many scientific protocols for review that it will reject after the first round of discussion because they place too high a risk burden on the participants. Other proposals will be rejected after the second round of discussion because the study sponsor, product manufacturer, or principal investigator failed to modify the study to the recommendations the IRB made in the first round of review. The IRB's rejecting a study does not mean that the study will cease to exist, however. Often, the study will be submitted—unchanged—to another IRB, as if that IRB were the first to consider it.

Each IRB member must become familiar with the basic concepts of the protection of participants of research and the extra protections and safeguards that must be extended to vulnerable participants. The IRB is considering not just research in general, but research involving humans, with risks of morbidity (disease) and mortality (death) to the volunteer, who may or may not possibly benefit from research.

Although the notion of balancing competing claims of the importance of a particular study is easy to advocate, its actualization can be exceedingly difficult. For example, say an IRB with a quorum present including the nonscientist (but three members absent) agrees that it has balanced competing claims to its satisfaction and votes to approve a research study. Had the three missing members been present, a different discussion might have taken place, different points illuminated and deliberated, and a vote in the opposite direction might have resulted. This example illustrates the weakness in quorum-based voting instead of requiring a vote of the full IRB membership for approval of a study proposal.

Although the Belmont Report states that "scientific research has produced substantial social benefits," the report notes that scientific research "has also posed some troubling ethical questions."[13] Thus, the IRB must focus all of its attention on the specific scientific hypothesis being put forth by the principal investigator and must ask itself what the benefit to society is from the research proposal and whether that benefit is worth the anticipated—and any unanticipated—risks of the study. There is a shortage of specific guidance regarding what methods can be used in balancing the pros and cons in the scientific and ethical debates that go on over each scientific protocol and informed consent form submitted to an IRB for review and consideration. Codes, as general guidelines, often do not specify what criteria should be used in weighing the risk to the study participant versus the benefit to society. This adds to the difficulty of deciding what is research and what is not. Although there is a trend toward considering more scientific activities and scientific proposals as research, and therefore subject to IRB review, it is not clear how widespread that trend is across the medical institutions in the United States today.

How Is Standard Therapy Different from Research?

New IRB members may not understand what is meant by *standard of care* and *standard therapy* (as in "What is the standard therapy for a particular medical condition?"). IRB members must understand the opinions on the standard therapies for a medical condition, because this will form the background of the board's reflections on a new drug or device that the study sponsor and principal investigator want to study. Standard therapy can include more treatment options than just drug therapy for some conditions; it can include dietary and herbal supplements and nondrug and nonsupplement modalities. Standard therapy also can include the possibility of observation of the medical condition without any treatment intervention. Unless an IRB member understands the standard therapy for the medical condition, he or she will not be able to understand why research on a new drug is needed.

Reasons a New Drug Is Developed

Sometimes it may be difficult for IRB members to understand why development of a new drug is being proposed. A new drug may be developed so that the manufacturer will have a product to compete with drugs already approved by the FDA and in use. A study may be proposed on a drug that is already approved for certain uses, but now the drug company seeks to expand the uses of the drug.

It is to be hoped that one goal of developing new drugs will be creating better therapies for the treatment of medical conditions. Often, an approved and marketed prescription drug may be effective in only 50 percent of patients with the condition the drug is prescribed to treat. The question a new drug study may be asking is whether a drug with a different mechanism of action will allow a higher percentage of patients to achieve the same benefit or an increase in benefit with equal or lower risk. A new drug may be judged to be better than standard therapy if it answers questions such as those in Box 2.4.

Box 2.4. Some Ways Improvement of Therapeutic Effectiveness Is Measured

- There is an increase in the number of patients benefiting from the new drug over standard therapy, measured in terms of increased survival, increased quality of life, or decreased risks of adverse side effects and other adverse outcomes.
- Even if there is no increase in the number of patients benefiting from the drug, the risk of serious side effects (e.g., death, stroke, damage to liver or kidney) are lower than with the standard therapy.
- The new drug has fewer serious side effects that affect quality of life (e.g., nausea or headache) than does the standard therapy.
- The new drug's positive effect continues longer than that of the standard therapy. (For example, the standard drug may have a positive benefit that lasts for 2 hours after one dose but the new drug provides a positive benefit for 4–6 hours.)

Why Do Research Studies Compare a New Drug to Placebo?

Aspects of the use of placebo in research are currently under debate in a variety of settings. Even when there are competitive drugs on the market, the *Code of Federal Regulations* allows drug companies to compare new drugs to placebo rather than to the competitors. There are advantages and disadvantages to this practice from various parties' points of view. I will discuss various issues about placebo use in later chapters.

Are Case Reports Research?

A case report is a report of a patient or a set of patients with a particular disease who have undergone a specific medical strategy in diagnosis, management, or treatment or have simply been observed over time with respect to a specific medical condition. If a physician or physicians believe that the case report may be a key contribution to the medical literature, they then summarize the data on the patients they have observed and cared for and submit this case report to the editors of a peer-reviewed journal for consideration for publication.

The IRB will have to ask itself whether a case report is research. The problem is that distinguishing case reports from research is artificial. For instance, if one bases the distinction on the number of individuals reported on, the questions arises, How many patients can be reported in a case report before it becomes research? Deciding that the number of patients (e.g., more than 2, 3, or 4) defines the difference runs into the difficulty that in the research literature there exists the concept of an N = 1 study. Thus, any distinction between case reports and research may not be reasonable.

If an IRB does not specify a number or does not specifically say that a case report is research that needs to be brought to the IRB for review, someone could assemble data on 1,000 patients and argue that this is a large case report, not a research study, and therefore not subject to the IRB's review. Therefore, the IRB should think seriously about having all case reports considered research and subject to IRB review.

Protection of Data Used in Research

One of the most important issues in medical research today is the need to preserve the confidentiality of participant data. This brings us back to the basic elements of informed consent as specified in the *Code of Federal Regulations.* The code specifies that the investigator, in seeking an individual's informed consent, must provide the individual with "a statement describing the extent, if any, to which confidentiality of records identifying the subject will be maintained."[14]

The IRB must continually emphasize the protection of research data in general and, in particular, research data that are related to a patient's clinical care or genetic data that involve not only the study participant but also his or her family.

Non-uniquely identified data are the easiest to protect, because there is no way to trace the data to an individual and his or her family. Even so, if, for instance, there was only one patient with a particular disease in a certain hospital or medical center at the time of the study, the identification of the disease would enable someone to identify the specific patient. The IRB must look beyond the simple inclusion of a name, Social Security number, or hospital record number to other ways in which study participants can be identified uniquely.

Not All Questionnaire Studies Are Nonproblematic

Some investigators think that questionnaire studies are easy to get through IRB review. The following example illustrates the problems questionnaire studies can get into when they raise highly emotionally charged issues.

A questionnaire study by a clinician or cognitive psychologist examining nonsensitive issues without unique identifiers may be considered at the lower end of the risk scale in its chances of causing harm to the participant. Highly emotionally charged questionnaire studies, however, may result in adverse outcomes, and the IRB must make certain the principal investigator is prepared to deal with such outcomes for the safety of the particular participant. Questionnaire studies with the potential for serious consequences for participants would include studies involving persons with post-traumatic stress disorder or depression or persons who are victims of violence. The IRB must be assured that the questionnaire study will be conducted in a setting in which there is access to highly trained counselors who can help any person who has a problem being approached to participate or actually participating in the study. In highly charged studies, these issues must be clearly spelled out as a risk of participation, so that an individual can decide if, when, and in what setting he or she is willing to participate.

What Does IRB Approval Mean to the Study Participant?

What does IRB approval mean in the mind of a study participant? At present, few are asking this question, but it should be part of the attempt to optimally protect the participants in a study. Until this issue is more fully studied, we can only ask ourselves further questions. Does IRB approval, in the mind of a potential volunteer, mean that the research proposed has been deemed worthwhile, expected to produce generalizable knowledge for future applications in patient care or further medical research? Or, at the opposite extreme, does a potential participant wonder if IRB approval means that the IRB thought the scientific protocol was "tolerable" as a scientific activity?

Do participants know that, although IRB members may be assigning a letter grade to scientific protocols and informed consent forms in evaluation of their quality—the quality of the science, the quality of the ethics, the quality of the informed consent form, the quality of the protection of participants—the final decision only reads "IRB approved"? A study that receives all Bs or mixed Bs and Cs and maybe a D in the review and evaluation procedure is passed with the same stamp of IRB approval as a study that receives all As.

Clearly, a concerted effort must be made to inform the public not only that the study was IRB approved but also that this approval should not reduce the seriousness of the question confronting an individual considering study participation: Is participation right for me? Principal investigators or members of the research team should not use the phrase "IRB approval" or "IRB approved" during informed consent sessions as if the decision made it okay for the individual to enroll in the research study. Individuals considering participation should never rely too much on the fact of IRB approval and shortcut their own deliberations over participating in research. They must do their own systematic reading and review of the informed consent form, asking questions whenever questions occur to them, and must conduct a self-analysis to make certain that all their questions are answered and they are certain that they want to bear the risks of study participation, understanding that severe adverse outcomes can occur even when not foreseen by the study designers and the research team. Individuals must determine, through whatever processes and with as much assistance as they need, whether they have the willingness and fortitude, not only to participate in the study but to tell the principal investigator if at any point they want to quit. In addition, the participant should ask the principal investigator who will pay the bills in the case of a severe adverse outcome, foreseeable or unforeseeable, directly caused by or indirectly related to participation in the study.

3 What Is Risk?

Risk in IRB work is often discussed in terms of two basic components: adverse outcome (harm to a subject), and the chance (probability) that adverse outcome will occur. Yet, much more is involved in risk than this simple framework suggests.[1] This chapter explores the many facets of risk that IRB members encounter.

Basic Concepts and Considerations

Even if they have no experience with research risk, IRB members may have had discussions of risk regarding clinical care for themselves or family members. When a physician recommends a diagnostic or therapeutic intervention, the patient weighs risk against potential benefit in deciding whether or not to pursue the treatment. The IRB member may have been presented with an informed consent form containing phrases that refer to harms or adverse outcomes that could occur with the recommended intervention. Typically, it would contain an estimate of the chance (probability) that an adverse outcome would occur. The estimate may be verbal (e.g., "rare chance of stroke or death"; "patient will frequently experience nausea and vomiting") or it may be numerical (e.g., "1 percent chance of stroke or death"; "20–40 percent of patients experience nausea and vomiting"). Because informed consent forms are used for all the individuals considering the intervention, the risks are described in terms of the mean or average patient and the data are obtained from the peer-reviewed medical literature and from the physician's, hospital's, and health care organization's experience across all of their medical facilities. This experience of risk contemplation from a patient's viewpoint is good to keep in mind, but much more awareness will be needed when doing IRB work.

Risk is a very complex topic, and IRB members must understand it in depth so that the IRB can address the complexities as it reviews and deliberates about research hypotheses, about the level of risk borne by study participants, and about how risk is portrayed (verbally, numerically, graphically, or otherwise) in scientific protocols and communicated in informed consent forms and informed consent sessions. The IRB deals with risks in key areas:

- the risks intrinsic to the drugs, devices, and other therapies that are developed and tested on humans to gain scientific knowledge for future populations

- the risk that is inherent in participating as a research subject
- the risks that harm may appear in participants, not during the course of the study, but some time after the study is completed
- the risk of conflict of interest on the part of study sponsors, principal investigators, medical institutions, or IRB members
- the risk of loss of privacy or confidentiality
- the risk that no compensation will be made by the study sponsor to individuals who sustain harm from participation in a study

Research Risk

The simple definition of *research risk* is the chance that a person will sustain harm or an adverse outcome as a result of participating in a study. The ten main aspects of research risk that IRBs must consider are listed in Box 3.1.

Box 3.1. Key Aspects of Research Risk

- the magnitude of the risk
- the severity of the harm that may occur
- the reversibility or irreversibility of the harm
- the temporal dimension of the risk (whether short, medium, or long term) and its duration
- what is known of the scientific and/or medical cause of the risk and why the risk occurs in certain individuals
- the perception of the risk by all concerned parties
- the cultural and linguistic problems of explaining the terms and concepts related to the risk
- the participant's understanding of what research is
- the risk to others from an individual's participation in research

Disclosing Risk to the IRB and to Participants

Institutional review boards deal with risk as it is portrayed in scientific protocols, in informed consent forms, and in advertisements recruiting individuals to participate in research. There is no guarantee that a full list of a proposed study's risks is provided to the IRB by the principal investigator and study sponsor. Therefore, IRBs must conduct their own searches of the peer-reviewed medical literature, identifying risks reported in related studies. For example, an IRB member can type the name of the study drug and the term *risks* into the search screen in PubMed and examine the abstracts available there. The IRB can also consult local and/or national experts about risks that a study they are reviewing may pose. In a sense, IRBs must be constantly tutoring themselves on how best to access information from electronic or expert sources, to ensure that the informed consent form contains all known risks and reasonable estimates of as yet unknown risks that may occur with the study drug or device.

The risks to individuals being asked to participate in a research study are disclosed primarily in the informed consent form, which lists the anticipated risks along with their estimated chance of occurrence. Risks are typically described as symptoms or signs an individual may experience (e.g., nausea, headache) or in terms of possible end results (e.g., death, stroke, heart attack).

Full disclosure refers to completely informing an individual of all the known or estimated risks that may occur with an agent or medical device that is the object of research. Terms like *foreseeable* or *reasonably foreseeable* are commonly used in the context of estimating risk. *Foreseeable risk* means that a harm may occur at some future time. For a *reasonably foreseeable risk* there is current scientific information suggesting a known probability and severity of harm. There can be multiple interpretations of whether a harm was or was not reasonably foreseeable before it occurred.

In IRB work, one must focus on clear definitions to prevent mistakes in interpretation of what is being said, why it is being said, who is saying it, what will be listened to and remembered, what will be forgotten, what is a correct interpretation of a term, and how a term may be misinterpreted.

Risk and Causation

Although the federal regulations do not address specification of the cause of an adverse outcome, (e.g., why the death or stroke or heart attack occurred in a study participant), an informed consent form should record how these outcomes *could* occur during the research study, as well as related information, such as how to recognize the onset of a stroke or heart attack and instructions to contact the research team or the medical institution or to call 911 if a problem occurs.

Estimated Risk

The risk of an adverse outcome or harm to a study participant is always estimated; it cannot be specified with certainty. The estimate may by derived from the peer-reviewed medical literature, by experts in science and medicine, or by physicians and scientists who are not experts.

Minimal Risk

A short survey or questionnaire study that asks individuals only for opinions is sometimes said to entail minimal risk. However, these surveys or studies can evoke memories and emotions to which an individual may have strong negative reactions, especially if an individual has had traumatic past experiences (e.g., child abuse, elder abuse, spousal abuse, post-traumatic stress disorder). These surveys and questionnaires thus have the potential for significant risk.

Federal regulations make assumptions about "risks encountered in daily life." The problem with this is that all participants may not share the same set of risks that are encountered in daily life. In addition, the IRB must consider not only the average or mean experiences with risk among participants but also the extremes of risk experienced by some.

Reasonable Risk

Federal regulations make the assumption that some risks to study participants are reasonable. But this assumption is based on the participant's understanding the following facts: that they may not benefit in any way from participation, that they are bearing risks without any assumption of benefit, and that if harm occurs they may not be compensated by the study sponsor and may need to take the case to court to get compensation. The participant must understand that the purpose of research is to gain scientific knowledge for future populations, not the population under study. It will be up to each individual considering participation to decide whether he or she finds the study risks permitted by the IRB to be acceptable for himself or herself.

Perceived Risk

Individuals perceive risk differently. Factors that may influence how a person perceives risk include personality, beliefs and values, recent and long-past history of adverse event in the person's life, and history of adverse event among significant others.

Risk-Benefit Calculation

In IRB work, risk-benefit estimation is the balancing of possible harm and the possible benefit that could result from participation in a research study.

The Components of Research Risk

For the purposes of the function of an IRB, risk may be considered to have components:

1. the nature of the risk,
2. the magnitude of the risk,
3. the severity of the harm and whether it is reversible, partially reversible, or irreversible,
4. the chance that risk will materialize,
5. the weighing of risk by the individual participant, and
6. how well the risk is being estimated.

The Nature of the Risk

When different people consider the nature of risk, they think about different aspects of an adverse outcome. Some IRB members understand the phrase "the nature of the risk" to refer to the physiologic system involved in a study-related adverse outcome. Other members view a risk's nature in terms of the mechanism of action by which a study drug affects a participant's body to cause a negative outcome. Still others interpret the phrase in terms of how the participant's daily life will be changed by a study-related adverse outcome.

The nature of the risk is separated into two broad categories: risk to survival and risk to quality of life.

A Participant's Survival. If there is a chance of a participant's dying during the study, the participant must clearly understand that participation in the study may cause his or her death. In the case of drug studies, severe allergic reactions (anaphylactic reactions) and severe bone marrow, kidney, liver, or central nervous system reactions can cause death. Some of these severe reactions may occur immediately on exposure to the drug; others occur in the short, medium, or long term.

The IRB member must ask whether there is a risk of dying immediately or in the short, medium, or long term. A case in point would be a research study comparing surgery versus radiation therapy for a cancer. Surgery may have the worst risk in the short term, because of the higher chance of the patient's dying during or shortly after the surgery. There are acute risks of dying from the anesthesia during the surgery as well as from surgical complications. If the patient escapes the immediate or short-term risk from the surgery and anesthesia, the surgical intervention may offer a better long-term benefit (e.g., over the next five years) than radiation therapy or chemotherapy. Radiation therapy will usually present no chance of dying immediately but does have the risks of radiation injury and offers a lesser chance of long-term survival.

A Participant's Quality of Life. Sometimes a participant in a research study does not face the risk of death, but the quality of his or her life might be negatively affected by study participation. The negative effects can include discomfort or pain, diminished ability to carry on with work (such as the effect of a stroke on the ability to walk or to use an arm or hand), and difficulty thinking clearly in the completion of routine tasks of daily living, all of which the participant was able to do better before the occurrence of an adverse outcome that is considered to have been caused by participation in the research study.

Adverse Outcomes after the Study Is Completed. Adverse outcomes can befall study participants after the research study is completed. For example, some study drugs adversely affect the bone marrow, and these effects may not be seen in some participants until years after the study is completed. Therefore, the IRB must be attentive to the possible long-term effects of study drugs and make certain that the patient has safe follow-up after study completion.

Unanticipated Adverse Outcomes. Even though considerable thought may have gone into understanding a scientific protocol, some adverse outcomes will not be predicted by the principal investigator, the product manufacturer, or the IRB. Therefore, the IRB member must make certain that a mechanism is in place for the principal investigator to report new adverse outcomes to participants who are enrolled in a research study.

The point here is that new risks may arise during the study that were not predicted by the IRB, principal investigator, or study sponsor. Thus, the nature of risks in study participation is such that new risk disclosures may need to be made to participants during the course of the study.

The Temporal Nature of the Risk. The nature of a risk can be described in terms of whether the adverse outcomes that can occur would take place in the short, medium, or long term. However, the IRB should ask the principal investigator questions to elicit the definitions of these terms. In the assessment of risk of a particular surgery, for instance, would the short-term risk occur within the first 30 days after the surgery, the first 60 days afterward? Would a medium-term risk occur between one and four years after the surgery, when 90 percent of patients receiving the surgery can be expected to die by the fifth year after surgery anyway? Or does a medium-term risk occur between five years and twenty years after the surgery, when 90 percent of participants can be expected to die by the twenty-fifth year after surgery? Does a long-term risk occur at five years or twenty-five years? If an IRB member is in doubt, he or she should ask the principal investigator to explain the time-interval term being used.

The Magnitude of the Risk

The term *magnitude of the risk* is somewhat imprecise because magnitude can be assessed and measured in many different ways. It can mean magnitude in terms of "the greatness of rank or position of a risk within a set of risks." Within any set of risks, there will be variations in ordering of the risks by individuals. This is one of the key points in how risks are communicated to study participants. The order in which risks are described to a participant may change the way in which the risk is viewed. For example, the first risk disclosed may be assumed by the participant to be the most important risk in the study. In a study with many risks, a phenomenon exists called risk-consideration fatigue, in which the perception of the importance of risk is influenced by the number of risks.

In our discussion, we will use *magnitude of the risk* to mean "the chance of an adverse outcome compared to the chance of that outcome without the study." Here, the magnitude of risk is usually expressed as a percentage. The IRB must evaluate the magnitude of risk—(1) to the individual study participant (whether the participant is at low risk, mean or average risk, or high risk), (2) to the entire population of participants being studied, and (3) to society—as it decides whether to approve or reject a study.

The magnitude of risk to an individual with a certain medical condition who is participating in a new drug study, for example, is the increase in the chance of experiencing an adverse outcome by participating in this study compared to not participating and continuing to take the standard drugs used to treat the medical condition. The magnitude of risk to the study population is the number of anticipated adverse outcomes attributable to the new study drug compared to the number that would be expected if that population were treated with therapies used in standard care. The societal risk is the loss to society if the study is undertaken and completed successfully (i.e., has clear outcomes) compared to the loss to society if the study is not undertaken and no new data are available for analysis, interpretation, and furthering of science.

Risk can be described in words (verbal probability estimates), such as *probable* or *possible,* as well as numbers (numerical probability estimates); or in numbers, such as percentages (%) or odds (e.g., 1 out of 10, 1 out of 100). My research has shown that patients differ widely in the numerical meanings they assign to verbal probability terms.[2] For example, the numerical meanings patients assign to the term *possible* range from 0.001 to 100 percent. Therefore, if verbal probability terms are used, they must be accompanied by the best numerical estimate available.

In informed consent forms, the magnitude of risk is usually represented as a "percent chance" (for example, "There is a 1 percent chance of dying during this new drug study"). In some, the sentence provides both percent and odds ratio ("There is a 1 percent [1 in 100] chance of dying during this new drug study"), which is better. The best wording would read, "There is a 1 percent chance of dying during this new drug study. This means that 1 out of 100 participants can be expected to die for drug-related reasons during this new drug study." The IRB can, and often should, require that the principal investigator not only provide numbers but also define what the numbers mean. This is important because even though some participants may not want numbers for a variety of reasons (e.g., lack of experience in using numbers in daily life), verbal probability terms, such as *low risk* can have a wide range of meanings: Does it mean 1 out of 10, 1 out of 100, 1 out of 100,000? The IRB must insist that the best available numbers be specified and that the numbers be broken down and further explained.

While the magnitude of risk may be expressed in numbers, the significance of that percentage or ratio depends on point of view. In a new drug study, for example, a 1 in 10,000 chance of a participant's dying will be very significant to him or her and to his or her family and friends, but how does society view this 1 in 10,000 chance of death, and how does the IRB view it? What happens if the chance of death is higher, say, 1 in 1,000, or 1 in 100, or 1 in 10? How do society and the IRB view these higher-magnitude risks? What happens if the chance of death is lower than 1 in 10,000, say, 1 in 100,000 or 1 in 1,000,000? How do society and the IRB view these lower-magnitude risks?

Understanding the Numbers in Calculating Risk. Even if a principal investigator specifies a risk numerically, what do the IRB member and the participant understand by the numbers? IRB members must ask where the numbers come from. Are they derived from a careful and systematic search of the peer-reviewed scientific and medical literatures, or are they the best estimates the principal investigator can come up with? The sources of statistics can make a big difference in the risk magnitude one can calculate from them. For example, the risk of dying during an intervention involving a patient with a complex set of diseases may be lower in medical centers that deal with complex patients on a daily basis; the risk may be higher in a smaller institution in which the surgeon may operate on complex patients only once per month. Other factors that may have influenced the risk numbers of interventions in other studies are the degree of expertise of the physicians performing the research intervention, and the degree of sickness of the patients at the other institutions (a large tertiary care referral center may see sicker patients than a primary care center).

If an IRB member has doubts about the numbers on which a proposal's estimates of risk are based, the member should search the peer-reviewed medical literature to find out the ranges of risk numbers for studies involving similar interventions in various types of medical centers. Before an IRB member searches the Internet for numbers and what they mean, it is helpful to hold a discussion with the full IRB to determine what numbers the IRB is interested in seeing and how best to search for those numbers. Although IRB members can search independently, they should bring the details of their searches of the peer-reviewed scientific or other publications to the IRB so that the full board can examine the published study, its data, results, and interpretations (like those in letters to the editor and editorial opinions in the peer-reviewed medical literature).

Ideally, a search of the peer-reviewed medical and scientific literatures conducted by either a principal investigator and study sponsor or an IRB should specify the names of the reviewers, the date on which the search was completed, the search terms used, the sources searched (e.g., PubMed at www.PubMed.com), the number of journal citations retrieved in the search, the number of journal citations reviewed by the reviewers, the results of the search, any problems encountered in the review process, the availability of the journals in the institution's medical library, and the interpretation of the journal articles in the light of what the principal investigator and study sponsor wrote in the scientific protocol and informed consent form.

Although the above documentation of a review was described as ideal, at some level, even if not at this ideal level, the IRB needs to document the amount of effort that went into the review and the limitations on the review posed by difficulties accessing medical and scientific articles.

In assessing the validity of the magnitude of risk estimated in the proposal and informed consent form, IRB members must ask whether individuals recruited into this study will face more risk than the numbers represent because of their underlying medical conditions, the medications they are taking, or other characteristics in which they may differ from the subjects in the studies on which the risk numbers in the proposal are based. Problems with the accuracy of the magnitude of risk also can result if the numbers represent the *average* risk borne by participants (in previous studies for which results are available). If so, what are the numbers for higher-risk patients? If individuals will be enrolled in this study who have more medical conditions and are taking several prescription drugs already, then these participants may be at higher risk than is suggested by the risk numbers in the informed consent form.

Communicating the Numbers in Calculating Risk. There is considerable and mounting evidence that people use *simplifying strategies* or *simplifying heuristics* in decision making.[3] These heuristics are short-cuts or rules by which people process information when making a decision.[4] One such device is what is termed the *availability heuristic:* people may believe an event more likely to occur if that event is more familiar (i.e., is more available to their recall). For example, events that are written about in the news media may take on the appearance of occurring more frequently. A risk may start to be perceived as relatively common because it is discussed frequently, but a study of the literature may suggest that the risk is in fact unlikely (say, 1 in 1,000,000). Another example is the overestimates that occur when people are asked to guess the frequency of rare causes of death, whereas people underestimate the frequency of more common causes of death, such as stroke or certain cancers.

The problem with these short-cuts and simplifying strategies is that they can lead to bias in how the people interpret information.[5] They may influence IRB members' decision making when they review risks related to scientific protocols, and they may affect how potential participants understand informed consent forms. One suggested remedy for this problem is to provide data in multiple different formats or to ask people in which format they prefer to receive information.

Little is known about what formats of risk information individuals consider to be informative and what formats they consider too confusing to understand. Some individuals report that they do not understand risk expressed in numbers (e.g., fractions, percentages, or odds ratios). In my research, I have asked patients how they prefer to receive risk information from physicians in clinical care. About half of the patients have reported that they prefer that risk be described in words (such as *probable, possible,* or *rare*), and half preferred that risk be described in numbers (e.g., percentages).[6]

Although many patients may prefer not to be given statistical information in numbers, they often fail to appreciate the problems of interpreting what is meant by a verbal probability term, such as *rare, possible,* or *probable.* One individual may not recognize the difference in the numerical meanings that another person might assign to a probability term.[7] The solution is to present risk in both numbers and words, following the same approach as advocated for controlling bias in giving information in general. Within certain limits (and with the goal of never being confusing), the goal should be to present risk information in ways that are most useful to the individuals being recruited into a research study.

The Risk-Benefit Ratio. In assessing the magnitude of risk of a study, IRBs consider the risk-benefit ratio. Because the IRB's primary goal is protecting human subjects, it may have a very different take on the risk-benefit ratio than do the principal investigator and study sponsor. There are at least two senses of the term *risk-benefit ratio* in clinical research. First, there is the risk-benefit ratio of the whole scientific study: whether the study should be carried out on humans and, if so, which individuals should be excluded and which should be included. The IRB must make certain that the principal investigator or study sponsor is not excluding classes of people who could benefit from the study or, conversely, including classes of people who would be unlikely to benefit from the study. Second, there is the risk-benefit ratio for the individual deciding to participate in the study. Although it is hoped that individuals with several diseases and those taking several medications will be excluded from participation because of the problem of interpreting whether a benefit or adverse outcome that occurred was causally related to the study, there are no guarantees. A principal investigator and study sponsor—intentionally or not—may be using inappropriate inclusion and exclusion criteria.

Exclusion criteria are selected so that individuals who are at high risk for adverse outcomes will be excluded from participation, as will individuals who are taking several drugs or who have several medical conditions. If an adverse outcome were to occur in one of these individuals, it would be difficult to say with any degree of certainty if the event was caused, for example, by the study drug or by a drug-drug interaction or was a natural occurrence related to one of the individual's medical conditions. Thus, certain individuals are excluded from participation because the study would not be safe for them despite the best precautions taken, or because the study's results would be compromised.

Inappropriate *inclusion criteria* can be of two types: "too loose" or "too tight." Criteria that are too loose may bring into a study individuals who are not appropriate because of the already-existing burden of risk they bear from their disease processes. Criteria that are too tight may, for example, bring into a study only individuals of a certain sex for a study that is not sex specific if the medical institution has a higher number of patients of that sex on its rolls, which would make for easier recruitment. For a disease that is not sex specific, information should be collected on both sexes, to determine if any sex-specific issues remain unidentified because of understudy of the sex not being recruited.

All must aim for adherence to inclusion and exclusion criteria that allow the safe conduct of a study for participants with appropriate results for the scientific questions being asked. The IRB must make certain that the risk-benefit ratio of the study is considered both in terms of the importance of the study for the advancement of the medical sciences in the care of future patients and in terms of current participants. The IRB must also make certain that the principal investigator and research team are making the appropriate choices in the selection of individuals to be included in the study.

IRBs sometimes use risk-benefit ratios inappropriately. For example, in deliberations about the amount of risk to be borne by participants relative to the amount of benefit "expected" from a research study, an IRB member might argue that the question is so important clinically and scientifically that a less-than-optimally designed study should be approved because some answer to this important question is better than no answer. Other IRB members would correctly argue against the proposal in its current form if the goal of designing a study that optimizes participant safety has not been met.

The Severity of the Risk

One way of examining risk is to ask how severely disabled the participant will be if an adverse outcome should occur. Will the participant recover completely without any problems, or will he or she have lifelong problems requiring medical care?

In assessing the severity of the risk involved in a study, the IRB must recognize that acceptance of any level of risk is the individual participant's decision. A person's past and current experiences related to his or her medical conditions, disease processes, past and current care, and past and recent research experiences may give a potential participant a dramatically different view of the severity of risk from that of an expert, the principal investigator, or IRB member.

The balancing of risk against benefit also influences the perception of the severity of the risk. For example, imagine a research study in which there is a 1 in 1,000,000 chance of a participant's death but the study results may demonstrate that the intervention being studied can extend the life of all people with

Box 3.2. Type and Severity of Adverse Outcomes

- Acute, emergency (e.g., participant ends up in an emergency room without any access to information about his or her participation in the research study)
- Reversible: Over what time period (e.g., will the reversible outcome last for 30 days, 1 year, 5 years)?
- Irreversible: With what level of damage to the individual (e.g., irreversible damage to liver or kidneys may shorten the participant's life)?
- Contraindicated (e.g., potential damage to the fetus if participant is pregnant)
- Uncomfortable or distressing (e.g., persistent diarrhea, nausea, vomiting, or headache)

a particular disease by one week. How would individuals asked to participate in this study react to this risk-benefit ratio in which the severity of the adverse outcome is extremely high (death) but the chance of the adverse outcome's occurring is extremely low (1 in 1,000,000) and the benefit to patients very small? Now suppose the chance of death is changed to 1 in 100,000 and the life extension to one month. With each incremental change in the chance of death and the life extension, there may be a change in one's evaluation of the severity of risk and threshold of willingness to support the risk.

The Chance that Risk Will Materialize

The chance of harm depends on a number of factors and is a more precarious number than it may first appear in a scientific protocol or informed consent form, which might refer to "about a 10 percent chance of an adverse outcome occurring."

First, has the particular intervention that will be used been studied and reported on in the peer-reviewed medical literature, and have the risks been reported in a substantial way? Often, because of the brevity of a scientific article, the risks may not have been reported in a substantial way. In addition, although risks of death may have been reported accurately, the risks that affect the quality of life may not have been attended to in the report. Even in the case of mortality and survival data, time intervals of interest to study participants may not be available because the investigators reported data only at years 0 and 5 and not beyond year 5. Investigators make assumptions about what data should be reported, and these assumptions do not necessarily accord with what IRB members and study participants want. For example, in past research studies on lung cancer and how different types of lung cancer behave under different therapeutic modalities (e.g., surgery versus radiation), the peer-reviewed medical literature focused on issues of survival at years 0 and 5, that is, the number of patients alive and dead at the time of treatment and the number of patients alive and dead five years after treatment. The number of patients alive and the quality of their lives at other time points will not be available if they were not sought and reported in these early studies.

Second, for a new intervention (e.g., new drug or new diagnostic or treatment procedure), there may be no available risk data, and thus the data provided to study participants will have to be the estimates of the principal investigator and other research and clinical experts in the field. How well are these estimates made, and how secure should a participant and the IRB feel about the estimates? (This question is discussed below.)

Third, in any research study, the inclusion and exclusion criteria used will limit the types of participant studied. For example, in a new drug study, individuals who are taking other drugs for medical conditions or who have several medical conditions may be excluded from participation. (Consequently, when the data from that study are published in the peer-reviewed medical literature, they will not address how the new drug will perform in the more complex patients who will be prescribed the drug once it is on the market.)

The Weighing of Risk by the IRB

As the IRB member reviews a scientific protocol, he or she should consider how a reasonable person would gauge the risks to survival and risks to quality of life posed by the study should an adverse outcome materialize in a study volunteer. It is important for the IRB member to assess how he or she might view the same issues if asked to participate in the study. Take, for example, the case of a stroke. Some individuals may consider potentially massive cognitive and motor losses a risk that they are unwilling to take, regardless of any possible benefit. The IRB member must clarify himself or herself how he or she views the risks of an intervention as contrasted with the wide range of ways individuals considering participation may view them. That is why it is important to both *specify* and *clarify* risks to the greatest extent possible. Then the member must weigh whether the scientific benefits of the research are significant enough to warrant study in human participants. The IRB must also attend to whether the risks borne by participants—with their varied medical conditions and burdens of disease—are justly distributed over all individuals bearing the risk in that population.

Usually, IRB members will need to consider three questions:

1. Is the scientific question being asked of sufficient weight and importance to justify the risks borne by the individuals and the group?
2. Has the study been designed in such a way that the risks borne by the individuals and the group have been minimized sufficiently?
3. Can these ideas about the risk be communicated to all study participants in an understandable fashion?

The IRB is instructed to weigh risk and benefit, if possible. However, study participants should be allowed to evaluate risk without any undue emphasis on unknown benefit.

The Weighing of Risk by the Individual Participant

The weighing of risk by individual participants most often starts with risk disclosure in the informed consent form. In clinical research, as opposed to clinical care, the major consideration will be related to risk, as research itself is an activity designed to benefit future generations, not the participant.

The individual is evaluating the risk using his or her values and understanding of the nature of risk in his or her life and the degree of risk he or she is willing to bear. The person may want to take the informed consent form home to discuss it with family, friends, or others. He or she may want to obtain the opinion of a primary care provider. Discussion with others should be encouraged, because it may improve the individual's understanding of the risk involved in the study. It also demonstrates that the principal investigator is sincere regarding the obligations of obtaining informed consent.

Individuals differ in their willingness to assume risk, the magnitude of risk they will tolerate, and their perceptions of the severity of risks and the value of benefits. The IRB must assess how reasonable the risks are in a study according to a range of personal preferences that participants might have regarding the risks. This is particularly important for studies in which the participant's decision-making capacity might change during the study (for example, in a new drug study in which a toxic effect of the drug may be manifested in mental and physical states that impair cognitive capacity).

There may be much more subtle changes occurring in an individual which may affect his or her capacity to weigh participation versus nonparticipation, continued participation versus dropping out of the study, or risk versus potential benefit. Very subtle reasoning and appreciation of risk may be needed to make these comparisons.

How Well the Risk Is Being Estimated

Much of the deliberation about risk by all parties involved in research is based on human estimation. Risk measurement in one medical institution, by one physician, by one principal investigator, may be radically different from that in another institution with other personnel. Also, people differ in their ability to estimate risk; some individuals' estimates always seem accurate, while others' are widely off track, underestimating in one circumstance and overestimating in another.

The IRB must ensure that individuals considering study participation have the best estimate of the chance of risk that can be developed from the data of earlier studies. However, risk has not always been a focus of the peer-reviewed medical literature; the reporting of discoveries of potential benefit has taken precedence over the reporting of failed studies. Thus, although the peer-reviewed medical literature remains a principal source for researching risks that have been reported and their frequency and severity, the IRB must also consult experts to estimate where the risks of participation lie in each study. Risk estimates must be calculated not only for the average or mean study participant but also for the medium- and high-risk participant, that is, the individual who is already at increased risk because of age, debility, physical or mental condition, or other factor.

Risk estimation is the developing of best guesses of the size or chance of the risk that an adverse outcome will occur in a particular participant or in a group of participants. Even if the risk numbers have been derived from the peer-reviewed medical literature, the search of that literature must have been complete and done systematically, so that the estimate can withstand scrutiny by an IRB. An IRB member should be able to search the peer-reviewed medical literature and come up with the same risk estimate numbers as the principal investigator. If not, then the IRB member must bring any discrepancies to the attention of the entire IRB.

Some IRB members argue that searching the medical literature should be solely the responsibility of the principal investigator and study sponsor. The fact of the matter is that IRBs often receive from principal investigators and study sponsors inadequately researched materials for review and evaluation. Inadequate searches of the peer-reviewed literature that were not challenged have resulted in the occurrence of severe adverse outcomes, including deaths, that could have been prevented. Thus, IRBs must take on the role of double-checking the reviews conducted by the principal investigator and study sponsor and must ensure that the full range of risks that could occur in the study is disclosed to individuals considering participation. Double-checking the principal investigator's and study sponsor's literature reviews is the only way the IRB can assure itself that all risks have been reviewed and are discussed. The task of searching the peer-reviewed medical literature may be daunting; a new IRB member should be paired with a more senior IRB member and nonscientists with scientists, and the institution's library staff can be helpful, so that IRB members can learn how to search the medical literature.

The IRB must be particularly attentive to severe adverse outcomes related to study participation, as principal investigators and study sponsors often downplay these, because of their low chance of occurrence and out of fear of disapproval of the proposal. Even brief checking of the peer-reviewed medical literature may reveal severe adverse outcomes that are known to be associated with the study intervention and yet are not disclosed in the informed consent form as presented to the IRB.

Box 3.3. Questions to Ask about Potential Harm to a Participant

- What would be the nature of the mental or physical harm that might occur?
- What would be the cause of the harm?
- What symptoms would the individual experience?
  —How severe?
  —How recurrent?
  —How permanent?
- What is the estimated chance of death?
- Would the harm affect the individual's life expectancy?
- Would the harm be to subject only or to others also (e.g., employability affected)?
- What social harms could occur (e.g., damage to social relationships)?
- How else might the individual's quality of life be different if the harm occurs?

Communication of Risk

Most of the information an IRB receives from a principal investigator is written. IRB members often face challenges in understanding what the principal investigator is trying to communicate to the IRB and to the potential study participant. The IRB must always emphasize the need for clear communication in understandable language.

Common Communication Themes in IRB Review

Certain themes permeate IRBs' reviews of how well scientific protocols and informed consent forms communicate necessary information. Principal investigators must attend to these concerns in their communications with potential study participants, and IRBs must attend to them in their communications with principal investigators.

Provide Clear and Precise Definitions. The job of the principal investigator is to translate the language of science so that scientific protocols can be understood by IRB members and informed consent forms can be understood by potential study participants. If a consent form is not clear to the IRB, how will it be clear to the participant?

Translate All Scientific Terms into Lay Language. The IRB member reviewing a scientific protocol must be able to understand the science of what is being said. If the IRB member cannot understand the science, even with the assistance of another IRB member, the scientific protocol should be returned to the principal investigator for translation into terms the IRB member can understand. Members cannot validly approve or disapprove a proposal they do not understand.

The IRB member reviewing an informed consent form must be certain that all scientific language has been translated into terms that a nonprofessional can understand. The IRB should try to help the principal investigator make such translations.

Avoid Minimizing Risk. The IRB must be certain that all foreseeable risks are disclosed to potential participants. However, the concept of foreseeable risk is subject to interpretation. Historically, principal investigators have been willing to disclose all common adverse outcomes but have been reluctant to disclose severe adverse outcomes with low probabilities. It is better to disclose all adverse outcomes, even those with low chances of occurrence.

Avoid Disclosing Too Little Information. Although the disclosure of a lot of information may overwhelm a participant, participants generally prefer to know the risks, and they cannot give a well-informed consent if pertinent information is not given to them. The information is being delivered in a written document that the participant can take home before signing. Many participants are quite resourceful and will check the information in the consent form against the *Physicians' Desk Reference* (PDR), the Internet, and other information sources. Who would want such a review to reveal that the principal investigator did not disclose information about risk that participants can easily find in their own search of the peer-reviewed medical literature? In a written form that is properly organized with clear headings, participants can select the information that they want to read and think about.

Avoid Using Exculpatory Language. Exculpatory language is wording that somehow clears a party from blame, obligation, or compensation related to an event that has occurred. The *Code of Federal Regulations* clearly states that there can be no use of exculpatory language in informed consents, whether oral or written.[8] According to the code, exculpatory language is language "through which the subject or the [subject's] representative is made to waive or appear to waive any of the subject's legal rights, or releases or appears to release the investigator, the sponsor, the institution, or its agents from liability for negligence."[9]

In an informed consent form, exculpatory language would be wording tending to say that the party in question is exempt from blame or legal obligation to compensate a study participant if harm occurs during or after participation that could be said to be related to the study. The participant always has a right to bring legal suit.

What Characterizes Poorly Written Communication?

The following problems characterize poorly written communication. Sometimes the problems lie not so much in the style of communication as in the poor design and structure of the study. That situation presents a different set of issues, which are discussed in Part 2 of this volume.

The "Nontranslation" Problem. IRBs often encounter scientific language for which the principal investigator has provided no clear definition. This occurs typically in the scientific protocol but can also be present in the informed consent form.

Box 3.4 shows two examples of the nontranslation problem as it might appear in the risk section of the scientific protocol and informed consent form for a study of a psychotropic drug. In Example 1, each of the scientific terms should be translated into language a nonprofessional would be likely to understand. However, simple translation is only the first step. The list of adverse outcomes and their chance of occurrence in Example 2 uses easier to understand terminology, yet it does not include the range of severity of each adverse outcome or the chance that an individual will experience more than one adverse outcome. For example, can the difficulty in breathing be so severe as to result in the need for intubation and placement on a ventilator? If so, how long might the intubation be required, and is there a chance that the person would not be able to be taken off the machine?

Box 3.4. Examples of the Nontranslation Problem

Example 1

Drug A has the following risks: cholestatic jaundice; fever with grippelike symptoms; leukopenia and agranulocytosis; motor restlessness of the dystonic type or resembling parkinsonism; dystonias with spasm of the neck muscles, progressing to torticollis; extensor rigidity of back muscles, sometimes progressing to opisthotonos; carpopedal spasm, trismus, oculogyric crisis, pseudoparkinsonism with masklike faces; drooling; tremors; pill-rolling motion; cogwheel rigidity, shuffling gait; tardive dyskinesia or tardive dystonia.

Example 2

Drug B has the following risks:

Incidence: Rare

- seizure or convulsion
- difficulty in breathing

- fast heartbeat
- high fever
- high or low blood pressure
- increased sweating
- loss of bladder control
- muscle spasm or jerking of all extremities
- pale skin
- severe muscle stiffness
- sudden loss of consciousness
- tiredness

Incidence: Difficult to specify

- lip smacking
- lip puckering
- puffing out of the cheek
- rapid or wormlike movements of tongue
- uncontrollable chewing movements
- uncontrollable movements of arms and legs

Failure to Separate Categories of Information. Separation of the scientific protocol into broad categories or types of information with clear headings is key in helping IRB members understand the document. Imagine asking an IRB of busy people to review a scientific protocol that has only one heading, "Study Protocol," followed by twenty single-spaced pages of tiny type. Box 3.5 suggests an outline by category of information that would facilitate review of the proposal.

Box 3.5. Categories by Which to Organize a Protocol

Research Team
- Principal Investigator, with Qualifications
- Co-Investigators, with Qualifications
- Clinicians Associated with the Study, with Qualifications
- Research Team, with Qualifications

Study Sponsor, with Qualifications

Conflicts of Interest, If Any, Related to Research Team or Study Sponsor

Study Title

Study Protocol
- Statement of Study Hypothesis

Background Materials Leading to Formulation of Study Hypothesis
Study Design
Number of Participants to Be Recruited
Sample Size Calculation for Why This Number of Study Participants Are Needed
Characteristics of Study Participants
Inclusion and Exclusion Criteria Used to Define Who Is Eligible for Study Participation (and Why) and Who Is Excluded from Study Participation (and Why)
Whether "Vulnerable Participants" Are to Be Included, and the Rationale for Inclusion
Procedure for Recruiting Participants
Risks of Study Participation
Discomfort of Study Participation
Costs to Study Participants
Liability, and Who Is Going to Pay for Medical Care, Hospitalization, Disabilities Related to Participation in the Study
Rights as a Research Participant
Persons to Contact with Research Questions Arising during Study Participation
Persons to Contact with Ethical Questions or Concerns Arising during Study Participation
How Research Data Will Be Stored to Protect Participant's Privacy and to Ensure Confidentiality
Compliance with HIPAA (Health Insurance Portability and Accountability Act), with Attention to the 18 Unique Identifiers Outlined by HIPAA Regulations

Blurring Distinctions. A presentation may blur distinctions by using overly complex terms or concepts that can obscure meaning. In general, distinctions can be blurred in the following ways:

- using inclusive concepts instead of breaking down concepts into simpler terms,
- using highly specialized terms instead of simple terms,
- writing long sentences with cumbersome phrasing,
- using verbal probability terms (such as *rare*) instead of numbers (such as 1 in 100,000),
- failing to organize categories of information with headings, and

- embedding concepts from one category of information in another category.

Blurring of distinctions tends to occur in two areas: the inclusion and exclusion section of the scientific protocol (where the principal investigator describes the characteristics of individuals to be included in or excluded from participation in the study) and the risk section of the informed consent form.

The inclusion and exclusion section of the scientific protocol should clearly state the evidence the principal investigator is looking for to characterize the participant. For example, in the case of individuals with hepatitis, will the principal investigator look at liver function tests only or will a liver biopsy be required? What happens if a principal investigator does not clearly state what he or she will examine but uses only a broad phrase such as "patients with compensated liver disease" to describe appropriate participants?

The risk section of the informed consent form should list in clear terms all risks participants might encounter. It is important to recognize that risk is a problematic concept, because it has many dimensions. The IRB member must ensure that potential participants understand health and medical outcomes discussed in an informed consent form in terms of what may actually happen. For example, if an informed consent form states that "the human participant may have an anaphylactic reaction," this may mean almost nothing to a layperson. In fact, a severe anaphylactic reaction may cause the participant to die or may require intensive care in a cardiology unit or intensive care unit. The participant may have to be intubated and placed on a ventilator to give the best possible treatment for this severe reaction. All of these ramifications of the possible adverse outcome need to be communicated to the potential study subject.

An IRB can never be certain whether blurring of distinctions was accidental or intentional. It may be an attempt on the part of the principal investigator to confuse less experienced IRB members, in hope of a quick approval, but actually such blurring causes problems in reviewing scientific protocols and informed consent forms and thereby delays the evaluation.

Blurring distinctions leaves ambiguity in the minds of IRB members reviewing the scientific protocol and informed consent form. The ambiguity must be resolved by having the principal investigator rewrite the materials, break down concepts into component parts, and reword the documents to make them clear.

Translation of Risk from One Language to Another

When the spoken language of the principal investigator or the person conducting interviews is not the spoken language of the individual considering study participation, the scientific language issues become embedded in a natural language problem. How does an English-only-speaking principal investigator discuss the issue and questions of the informed consent form with someone who speaks only Spanish, Chinese, or Japanese?

Real solutions lie in correct translation of informed consent forms and the employment of individuals who speak fluently the language of the participant to conduct the informed consent session.

Cultural Aspects of Risk Communication

Even with explicit translation, differences in culture may block the understanding of scientific and medical terms for individuals considering study participation.

The Need for Repeated Review and Clarification

At the end of the initial review of a proposal, the IRB will often make recommendations involving clarification of language. Sometimes, IRB members will be able to suggest how to translate scientific or technical language into nonscientific language.

If the clarifications needed are simple, the principal investigator can probably make them easily. However, if many changes are required or the subject matter is in a research or ethical area in which the IRB has not had previous experience, then the principal investigator may have to work hard to translate the new material into language a nonscientist can understand.

With a complex scientific protocol and informed consent form, it may be difficult to identify all the necessary changes in the initial review. When the scientific protocol and informed consent form are resubmitted in revised form, they should again undergo extensive review. The IRB may identify issues in the revised version that were not evident in the original. As the IRB better understands what the principal investigator is talking about, new questions may be generated, and these new questions will require further changes and further IRB review.

Often, the principal investigator feels that his or her only task is to correct the "blurred distinction" and then the project will be approved. This is frequently not the case. The principal investigator's clarification of one blurred distinction may reveal deeper problems that require much more work on the part of the principal investigator to yield a study protocol that optimally protects the participants.

The IRB may ask the principal investigator to come to a full board meeting to discuss issues or may designate an IRB member to discuss issues with the principal investigator. These meetings will result in written changes made by the principal investigator which must then be resubmitted to the IRB.

Full Disclosure of Risk

Informed consent requires full disclosure of information in a written document, the informed consent form. Before signing the form, potential participants can take it with them to consider, to generate questions for the principal investigator or IRB chair, and to discuss with others.

What Is Full Disclosure of Risk?

Each IRB must decide what constitutes full disclosure of risk for each study. Although it is clear that all severe adverse outcomes, even at low probability of occurrence, must be disclosed, the IRB must consider the full range of severities of adverse outcomes and the full range of probabilities. When in doubt, I recommend clear and complete disclosure.

The Chance of Dying, and What Can Cause Death

Some informed consent forms may state that a participant in the research study has a "chance of dying," and others express a numerical probability estimate of dying (e.g., 1%). Because some study participants may be completely unfamiliar with the term *probability*, its use in an informed consent form may cause unnecessary confusion.

Where possible, an informed consent form should also say what might cause a participant to die during the course of the study. IRBs vary in their approaches to the disclosure of causation. Take the example of three IRBs reviewing a study in which there is a risk of dying from cancer chemotherapy. IRB 1 insists that the principal investigator state that the participant may die in the study. IRB 2 insists that the principal investigator provide the best reasonable estimate that any participant may die in the study. IRB 3 insists that the principal investigator state (a) that the participant may die; (b) the chance of any participant's dying, depicted verbally and numerically; and (c) the possible causes of death related to study participation. This study is comparing two cancer treatments, so the possible causes of death might include:

- infection due to the low white blood cell count produced by the drugs' effects on the bone marrow
- bleeding due to the low platelet count produced by the drugs' effects on the bone marrow
- anemia due to the low red blood cell count produced by the drugs' effects on the bone marrow

The IRB must determine which approach is best for individuals' understanding of the chance of death from participation in that particular research study.

The Development of Risk Criteria

To optimally protect study participants, an IRB member must understand the criteria used in the consideration of risk. Federal regulations are of limited help in clarifying risk concepts. Some regulations are specific (e.g., each IRB must have as a member of its board at least one nonscientist), while others are general (e.g., vulnerable participants need special protection) and are really only guidelines. In many instances, an IRB must develop its own criteria for applying the guidelines to a research proposal. The risk criteria it devises should be discussed with regulators to verify their consistency with the regulations.

Criteria for the Level of Risk

The levels of risk are easy to label, but those labels are sometimes hard to define and distinguish from one another. IRB members should consider the following questions and what numerical meanings they would assign to each of these risk phrases: What is "a low risk," "a high risk," "minimal risk," "possible harm," "probable harm," "definite harm"? How do the numerical equivalents of the verbal phrases answer the question "How many participants would be expected to sustain an adverse outcome during the study?" Would the participant give the same numerical meaning to the terms as the study sponsor and principal investigator are doing?

Often risks and severity are difficult to verbalize and describe in operational terms. For example, death and stroke are very severe adverse outcomes, but what about congestive heart disease or depression caused by a new experimental drug? Whereas some individuals may consider certain common side effects of medications (e.g., nausea, constipation) to be of less concern, other individuals may consider these side effects to be of great concern in terms of the experience in their lives. Some individuals may weigh the risk of blindness differently from the risk of paralysis; other individuals may weigh them equally. Despite this diversity of perception, the IRB will have to determine what criteria will constitute which level of risk for a particular study.

In an informed consent form, it may be simpler to include just the name of the adverse outcome, its degree of severity, and the chance of its occurrence and not to address the category of risk the adverse outcome falls into. But, in its oversight of the study, the IRB will have to find a way to categorize the level of risk in the study. This categorizing of levels of risk should correlate with the intensity and frequency with which the IRB will perform ongoing monitoring of studies. The higher the level of risk, the closer the monitoring will need to be.

Criteria for Assigning Numerical Risk Levels to Severe Adverse Outcomes

One approach to assigning criteria to severe adverse outcomes is first to determine what the potential severe adverse outcomes are (e.g., death, stroke, paralysis) and then to assign a numerical level of probability (chance) that each outcome will occur. An IRB can use this approach to begin discussing the criteria it wants to assign to both specific adverse outcomes and their probability of occurring.

Criteria for Assessing Risk and Benefit

The IRB must recognize two types of research: studies in which there is a chance that a participant may benefit, and studies in which there is no chance that a participant may benefit.

An example of research in which there is a chance that a participant may benefit is a study of a new lipid-lowering agent. In certain categories of patient, the lowering of lipids will be a benefit that can be measured and compared to the chance of harm (e.g., irreversible damage to the liver because of the study drug). One might determine that there is a 70 percent chance that the study drug will lower the participant's cholesterol with a 0.01 percent chance of irreversibly damaging the liver of an individual without previous liver damage and without risk factors for damaging the liver.

An example of research in which there is no chance of benefit to the participant is a genetic study in which an investigator is attempting to identify genes or gene sequences whose presence indicate that the participant will develop an untreatable neurological condition. The participant cannot benefit medically from the study, but the eventual neurological damage will not have been caused by the study. The research project may continue after the participant has died of the neurological disease, but the death will not be an adverse outcome of the study.

Comparing studies on both ends of a risk-benefit scale (minimal-risk observation studies versus high-risk invasive studies) can help an IRB determine how it will categorize the risks and benefits involved and how these risks and benefits will be explained to the individuals being recruited to participate. The IRB can then track its choice of criteria and categorization over time to see how well they have worked and how the board might improve in its never-ending quest to do a better job.

Part II

The Scientific Protocol and the Informed Consent Form

4 Prescreening of Proposals

The IRB must be attentive to both the quality of the materials the principal investigator submits for review and evaluation and the qualifications of the principal investigator and the research team who would carry out the study.

The Quality of the Submitted Materials

There is a wide range of quality in the materials that are submitted to an IRB for review and consideration for approval, modification, or rejection. Some proposals are well thought out and others are hastily put together. Some look as if minimal thought has gone into their most basic aspects: how the scientific hypothesis is formulated, how the science would be executed in the study's methodology, and how human participants would be protected during the course of the study.

An example of a substantial omission in a principal investigator's plan is proposing a study of a treatment without developing a strategy for monitoring the laboratory tests and study reports that would be done on individuals before they enrolled and during the study. Another substantial omission is the failure to specify qualified individuals who will detect and act on laboratory and study abnormalities, communicate with both the participant and the participant's primary care provider about abnormalities, and ensure that the participant is cared for appropriately. These qualified individuals should be in place before the study begins, because abnormalities may be detected before study entry, during the study, or after the study.

Some individuals who hope to become principal investigators do not have the skills to develop a scientific hypothesis, design a research study, or write an informed consent form. The institution should provide instructive programs for young faculty members, to help them develop the necessary skills and to provide models. If the institution does not have such programs in place, the IRB will often see incomplete proposals.

It is to be hoped that the IRB, with the help of the research staff of the institution, can set up a plan for prescreening all studies to detect deficiencies in the science, in the protection of participants, or in the wording and formatting of documents. This prescreening can be done with the help of the IRB chair or its members in conjunction with the IRB's staff.

Careful prescreening by qualified IRB personnel minimizes wasted time when the IRB meets as a full committee. The more the IRB anticipates problems and recognizes ambiguities early on, the more it can provide guidelines to principal investigators and study sponsors on how to generate well-defined, clearly developed, and unambiguously worded materials. If ambiguities remain in the wording of the documents, the IRB must spend time asking the principal investigator basic questions that were unanswered because of lack of work or skill by the principal investigator or study sponsor.

Prescreening can also reduce the time taken to correct grammatical errors, translate medical terminology into language a study participant can understand, catch numbers that do not add up, and other details. If prescreening works well, the full IRB when it meets can focus on its primary role of evaluating the science and ethics of the proposed study. If no prescreening is done or if the prescreening is inefficient, most of the IRB's time may be spent on preparing the materials to go back to the principal investigator to eliminate the deficiencies so that the IRB can then deliberate about the more meaningful issues of the study.

Prescreening involves three phases. First, there is prescreening of the research team. The task is to identify the members of the research team and determine whether they are qualified to perform the study. This is particularly important in studies that cross services, such as an internist proposing a study involving psychiatric patients or a psychiatrist proposing a study involving medical patients. In my opinion, a principal investigator should not conduct a study in an area outside his or her expertise without involving a co-investigator or collaborator from that discipline or service. The IRB must be certain that there is a co-investigator or collaborator from the medical service in which the principal investigator does not have clinical expertise. Clinical expertise on the research team is important for participant safety and for the communication across those services that will be needed to deal with any medical problems that may arise during the study. For example, if a psychiatrist is proposing to study individuals with diabetes, who will be watching the participant's diabetes, checking the prescreening laboratory tests and studies, and monitoring the follow-up of the laboratory tests and studies? Who will be communicating with the participant and his or her clinician regarding the study? The best person to communicate with a participant's primary care physician about test results is one who understands them well. The qualifications of the research team are discussed in greater detail later in this chapter.

The second phase is checking to see that all the key components of the scientific protocol are present (see Box 4.1). If they are not present on the initial

Box 4.1. Key Components of the Scientific Protocol

- Scientific objective (the scientific hypothesis to be tested)
- History of research relevant to the topic, including adverse outcomes occurring in similar studies
- Feasibility of completing the study with minimal risk to participants
- Scientific methods to be followed
- Criteria by which an individual will qualify to participate in the study (inclusion criteria) and by which an individual will be disqualified from participation in the study (exclusion criteria)
- Experimental laboratory tests and studies (versus those that may be part of clinical care) that will be included in the study
- Presence of a data safety monitoring board
- Sample size (the number of participants to be included in the study)
- Methods of analysis of results (statistical analyses or other techniques that will be used to analyze the study data)

submission, then the prescreeners decide whether to return the scientific protocol to the principal investigator for revision before the full IRB evaluates it or let the proposal be reviewed in its present form and then again after revision. Chapter 5 discusses in depth the review of scientific protocols.

The third phase is checking to see that all the key components of the informed consent form are present (see Box 4.2).[1] Omission of any element will cause the IRB problems in understanding and evaluating the study. Therefore, identifying any omissions through prescreening will save the IRB time in its initial review and consideration. As with the scientific protocol, the informed consent form can be sent back to the principal investigator for revision before the IRB considers it or the IRB can consider it with the omissions and then re-review it after it is revised. Chapter 6 discusses the review of informed consent forms.

The question of who does the prescreening is important. Experienced study coordinators with access to the IRB chair or other experienced, designated IRB members will probably be involved in the prescreening process because they have the expertise necessary to deal effectively with the issues that come up in reviewing a scientific protocol and informed consent form. The IRB will have to deal with any problems that have not been resolved in the prescreening process.

Box 4.2. Key Components of the Informed Consent Form

- Statement that the project is "research"
- Statement that research is not clinical care, although clinical care may be provided during the study (e.g., when an adverse outcome occurs)
- Description of the scientific purpose and objective of the study
- The procedures to be followed by the participant, including times and days (perhaps best conveyed with a timeline or flow chart)
- A clear delineation between the experimental procedures the participant would undergo in the study and the clinical procedures the individual would undergo for the medical condition if he or she were not in the study
- Clear statements about the risks of participating in the study. Risks should be described in terms of their nature, severity (including reversibility), and chance of occurrence.
- Clear statements that unforeseeable outcomes may occur during the study, despite the best efforts of the principal investigator to identify and prevent them
- A statement of any benefit that may accrue to the participant. Benefit should be described in terms of its nature and chance of occurrence. Furthermore, the quality and probability of benefit should never be overstated.
- Disclosure of any alternative procedures or treatments that might be advantageous to the participant. These alternatives would be available to the individual in routine medical care in the community if he or she decides not to participate in the study or after (but not during) study participation.
- Clear statements about how confidentiality of research records will be maintained and to what extent data will be kept anonymous.
- Explanations of: what will be covered in terms of hospital care if an adverse outcome occurs to the participant during the study; what compensation would consist of, if it exists; and where the participant can go to obtain further information about these two matters
- These three questions must be answered:
  —Who will answer a participant's questions about the research?
  —Who will answer questions about the rights of a participant?
  —Who will answer questions about research-related injury or death?
- Which member of the research team the participant should contact with any questions or concerns once enrolled in the research study
- Whom the participant can contact on the IRB with any questions about or problems with the way the research study is being conducted
- A statement that participation is voluntary and that the participant may

withdraw from the study at any time without giving a reason and without loss of any benefits he or she may be receiving from the institution apart from the study

- If the study will involve storage of uniquely identifiable tissue, there must be additional statements and additional informed consent forms. The participant must agree to the storage, understand where the tissue is to be stored, and understand his or her right to withdraw the tissue from use for research purposes and to have it destroyed.
- If compensation is to be involved, the type and amount and when and how the compensation will be paid must be specified.

Early Education and Training of Principal Investigators and Research Teams

A key adjunct to prescreening is early education of principal investigators and research teams in what is required for a high-quality submission to the IRB. Medical research institutions may consider detailed instruction and training for all new principal investigators and new members of research teams on their arrival at the institution. This early instruction includes a description of what constitutes a high-quality submission to the IRB and of the amount of work that needs to go into preparing one. Early training is a potentially powerful educational tool that will serve the institution and IRB well, yielding a higher-quality IRB review process for all scientific protocols and informed consent forms.

Continued Retraining

The training of researchers needs to be done more than once. Continued training at key intervals is important in sustaining a commitment to high-quality submissions to the IRB. Any change in or clarification of the regulations regarding study participants should serve as a stimulus to review what is needed in research proposals. The IRB study coordinators and chairs can keep track of the areas where the institution's principal investigators are excelling and opportunities for improvement and can report these results at all training and retraining sessions.

Communication and Resubmission

The IRB must be systematic in its communication with the principal investigator after reviewing the scientific protocol and informed consent form. The board should present all recommendations in a point-by-point format. The principal investigator's responses should use the same format and address all points that the IRB has identified as needing clarification or improvement. This process provides clear and precise documentation of the communication between the IRB and the principal investigator and allows both to keep track of the points that have been clarified and the work that needs to be done.

All resubmissions by the principal investigator to the IRB should be prescreened. The resubmitted materials must be checked to ensure that the principal investigator has responded to each point. Otherwise, the IRB's discussion will be incomplete and any basis of IRB decision making will be incomplete. All issues should be resolved in prescreening before the full IRB considers a resubmission.

The Qualifications of the Research Team

The qualifications of those conducting research have to be evaluated in both science and ethics. Regarding scientific expertise, the IRB needs to assess the qualifications of the principal investigator's research team not only within the principal investigator's own area of expertise but also across services if a principal investigator will be dealing with participants who have medical problems outside of his or her expertise. For example, if a study involves the use of a questionnaire to detect depression and depression is identified in a participant, how will the principal investigator ensure that it is treated promptly and properly? Or if a principal investigator will be enrolling participants with impaired or diminished decision-making capacity, how will the principal investigator assess competency and decision-making capacity? These issues must be documented, reviewed, and evaluated by the IRB to ensure that they have been addressed.

Box 4.3. Questions about Scientific Expertise Needed on a Research Team

- Does the principal investigator have the expertise necessary to take care of any clinical problem identified in a study participant?
- If not, do members of the research team have the needed expertise? If so, what specific qualifications do they have?
- Has the principal investigator identified relevant services in the medical institution that will be needed if a clinical problem is identified in a participant?
- What qualifications are needed in counselors for questionnaire studies involving domestic violence?

- What qualifications are needed in a health care provider who is responsible for identifying abnormalities in clinical laboratory tests or clinical studies involved in the study, and who is responsible for acting on those abnormalities to provide care to the participant in a timely fashion?

The ethical expertise of a research team primarily relates to their ability to carry out the research while maintaining as their primary aim the protection of participants at all phases of the study. Protecting participants must be the primary role of not only the principal investigator, who is the leader of the research team, but all members of the team. They must understand exactly what it means to protect participants; it takes constant vigilance by the research team, the study participants, and the participants' primary care physician or other subspecialty physician to ensure that all questions and issues related to participant safety are immediately identified, communicated, and acted on. (During a study, participants are usually also followed by their own health care provider in the medical institution and/or in the community. Indeed, if the participant travels during the study, the research team must have in place a plan to establish communication at any time if a question arises regarding the individual's research participation.)

The most difficult challenge of a medical institution's research service and IRB is continuously to train and remind the research team of their obligation to protect participants throughout the entire research process. This is particularly important for new principal investigators and research team members who have not previously worked in that institution. Differences exist among institutions about interpretation and application of the guidelines for the protection of participants set forth in the Belmont Report and the U.S. *Code of Federal Regulations.*

The protection of participants and scrutiny of research may be different in foreign countries from that required in the United States. The legal implementation and judicial interpretation of informed consent in clinical care and research are approached differently around the world. For both U.S. citizens doing research in other countries and foreign nationals doing research in the United States, it is important to find out about the host country's laws regarding the conduct of research on human subjects. As soon as any new principal investigator or new member of a research team in a medical institution appears for credentialing, the process of education concerning the protection of participants should begin.

5 The Scientific Protocol

The main task of the IRB is to protect human participants in research. To complete this task, the IRB member must be able to understand the science of what is being proposed in a research study and why it is being proposed, that is, whether a particular scientific question may be answered by studying humans.

A scientific protocol is a document that describes the scientific hypothesis being posed, provides a background of the science that has been completed on the hypothesis up to that point, and argues why this research project should be carried out and, if it is sponsored, funded by the study sponsor. This description of the scientific objective is followed by a description of the study methods to be used, why the data to be collected are crucial to the study, how the data will be analyzed and by whom, and the risks of the study to the study participants, to the environment, to the research staff, and other risks. The documentation of each step in the scientific method allows for replication of the results by future investigators, who might extend the original hypothesis in new directions.

The scientific protocol must elaborate how the study participant is to be protected optimally, in physical and mental well-being, at each phase of the study (and afterward for any study whose effects may be felt after the study is completed) and during the identification of and action on any abnormalities that are identified on entry into the study or that occur during the study. Participant safety must be addressed in the formulation of the study hypothesis, in the decision making by the principal investigator about which individuals will be included in or excluded from the study and why, and in decisions about which laboratory and study data will be collected and why.

A scientific protocol must present all details in such a manner that the IRB can easily grasp all ramifications of the study and possible adverse outcomes that may occur. All risks must be researched thoroughly, verified in up-to-date peer-reviewed medical literature, and clearly specified so that potential study participants can adquately consider whether they wish to enroll in the study.

In the scientific protocol, the principal investigator must present the latest scientific information about the reasonableness of the study hypothesis based on a thorough and systematic search of the peer-reviewed medical literature and a discussion with colleagues about the scientific worthiness of the hypothesis and the accuracy of the estimated risks of the study. For a complex study involving a drug with a new mechanism of action, an IRB may want a principal investigator to include endorsements from experts, commenting on the research proposal. The IRB must be able to recognize whether or not the literature cited is complete and up to date. If it is not, the IRB must do its own search of the literature and consult experts to determine whether the scientific protocol lists all known or suspected risks.

If a study is to involve human participants, the scientific protocol must ensure that the study is organized and structured so that there will be minimal risk of harm to participants and that the research team includes trained and experienced members who can recognize an adverse outcome at the earliest possible point and immediately communicate with the participants and act to minimize any damage to participants. This requires clear communication among the research team, the study participant, the team's clinicians, and the participant's clinicians to ensure optimal management and treatment.

Some IRBs may receive scientific protocols after a research committee has already reviewed them; in other institutions the IRB alone performs the review, analysis, and evaluation of the protocol. In either case, each IRB member must be prepared to review the science of each study. An outline-style check list for reviewing scientific protocols is provided as an appendix at the back of this book. IRB members must be able to answer the following questions about the study:

- What is the study objective or hypothesis?
- What is the key research question that this study is designed to test?
- Why is this study being proposed?
- Why is the question important enough to require study with human participants?

The Structure of the Scientific Protocol

The scientific protocol typically is divided into sections. It will probably contain the elements shown in Box 5.1.

The Study Objective/Scientific Hypothesis

The question, "What is the study objective?" has many alternative phrasings, for example, "What is the scientific hypothesis?" and "What scientific question is being asked?" In approaching the hypothesis, an IRB member should ask him- or herself:

- What is the scientific hypothesis (What scientific question is being asked)?
- Why is this question being asked?
- Why is this question being asked now?

Box 5.1. Elements of a Scientific Protocol

Basic Overview

1. Abstract
2. Scientific hypothesis to be tested
3. Rationale
4. Experimental design (approach)
5. Risks to participants
    - 5.1. How adverse outcomes are to be identified early
    - 5.2. How adverse outcomes are to be communicated to participants
    - 5.3. How adverse outcomes are to be managed with minimal harm to participants
    - 5.4. How participants are to be compensated if irreversible damage occurs

A More Detailed View

1. The nature of the study
    - 1.1. Hypothesis/study objective
    - 1.2. Rationale: why this study should be done in humans
2. Principal and other investigators, clinicians, and members of the research team
    - 2.1. Biographies
    - 2.2. Qualifications to conduct the proposed research
3. Study design (approach)
4. Participant population
    - 4.1. Total number authorized for this site
    - 4.2. Accessibility to all ethnicities and to both males and females
5. Risk-benefit ratio
    - 5.1. Beneficence
    - 5.2. Justice
    - 5.3. Respect for persons
6. Detailing of all risks (adverse outcomes)
    - 6.1. Estimated as foreseeable
    - 6.2. Estimated as unforeseeable
7. Benefits, if any
8. Vulnerable participants (including the extra protections to be given)
    - 8.1. Minors
    - 8.2. Fetuses
    - 8.3. Pregnant women
    - 8.4. Prisoners
    - 8.5. Mentally ill persons with impaired decision-making capacity
    - 8.6. Medically ill persons with impaired decision-making capacity

8.7. Other participants with impaired decision-making capacity
8.8. Others
9. Drugs/devices
9.1. Specific names of drugs and/or devices to be used in the study
9.2. Clear distinction between experimental interventions used in the study and interventions that the participant would receive as part of standard care whether or not he or she was in the study
9.3. Dosage of drugs used in the study
10. Data and safety monitoring
11. Anticipated endpoints/outcomes of the study
12. Statistical methods to be used in analyzing study results
13. References
13.1. History of the scientific question to be asked in the study hypothesis. Answer to, Why is this study hypothesis important enough to risk the occurrence of adverse outcomes in humans?
13.2. Background searches of the peer-reviewed medical literature (particularly related to risks and participant safety) and the completeness and depth with which these searches were conducted by the study sponsor and the principal investigator

- Why is this question being asked here (i.e., in this institution)?
- Does the institution have the scientific and ethical expertise to undertake this research?

The study objective and its scientific hypothesis should be understood in terms of the background of research relevant to the hypothesis, including adverse outcomes that occurred in similar studies, and the feasibility of completing the study with minimal risk to participants. Research has been described as "systematic investigation . . . designed to develop or contribute to generalizable knowledge."[1] Beyond satisfaction of that criterion, one can also ask: Is this scientific question new? Does the scientific question build on previous research? Are there other reasons why an investigator is pursuing the study?

"New research" is research in an area of discovery and exploration (for example, investigating a disease that is rare in the population and has not been investigated before). Research can be new but also build on previous research, and in the scientific protocol the principal investigator will describe and refer to relevant previous research reported in the scientific literature. Some research purposely duplicates previous research but can result in new outcomes, such as testing to see if a different medical institution and different physicians obtain the same results obtained at another institution. One medical center may use a new device and report on it in the peer-reviewed medical literature; other institutions may then decide to see if they can duplicate the results. Many new surgical techniques are developed and refined in this manner.

Sometimes the scientific hypothesis is not clearly stated. If the IRB reviewing a scientific protocol cannot determine what the study objective is, then the board must ask the principal investigator to clarify and restate the objective or hypothesis.

The Real Purpose of the Study

An IRB member must be cognizant of, or the IRB in its review may discover, intentions of the study not revealed in the protocol. For example, a study proposed by a drug manufacturer may have a legitimate research objective (e.g., establishing a new use for a drug approved for another use), or it may be proposing the study so that the company's drug will be introduced into use in the medical center where the study would happen. When a proposal is submitted, the IRB must decide whether the objective is pursuit of a better benefit profile or better risk profile for a treatment, or if the real objective is something else. The institution must decide if it wants to be involved with research studies with limited scientific value and possibly ulterior objectives. These are tough decisions, and many perspectives should be discussed at the interface of IRB work with the institution's research service, the institution's clinical services, and the institution as a whole. There may be a conflict of interest.

The introduction of a new drug or medical device into a medical center can cause problems. For instance, the health care providers at the institution will be caring for the study participant long after the study is completed, and it is they whom the patient will ask for the drug or the device after the study. Product manufacturers will not necessarily agree to supply the drug or device to an individual outside the study. As a responsibility to the institution's clinicians, the IRB needs to see that the manufacturer's post-study intentions are clarified. The IRB should ask the principal investigator what will happen regarding the study drug or device at the end of the study.

As a responsibility to study participants, the IRB should ask the principal investigator and study sponsor what will happen if a participant simply stops taking the study drug. This is particularly true of cardiovascular medications used to control potentially lethal arrhythmias, like ventricular tachycardia. In such a case, the IRB may need to insist that the consent form specify the risks of stopping the drug without the counsel and approval of the principal investigator. This is not to say that a study participant would not have the right to withdraw from the study whenever he or she wants to for whatever reason. Rather, it is a safety issue, because certain drugs cannot be stopped suddenly without disastrous consequences.

Basic Questions about the Scientific Protocol

Here, divided into categories, are the basic questions that should be asked about the scientific protocol.

The Nature of the Research

- What is the research being proposed (e.g., comparing a new drug to a standard drug, to a placebo)?
- Why is the research being proposed?
- Why is there a need for the research to be conducted on humans?

Principal and Other Investigators, Clinicians, and Members of the Research Team

- Who are the investigators and the members of the research team? What are their qualifications to conduct the study? Are they credentialed in the medical institution? What is their study experience?
- Who is in charge of early detection of adverse outcomes occurring during the study, communicating with study participants and their primary care or specialty physicians about adverse outcomes, and acting to minimize the occurrence of problems related to adverse outcomes?

The Objectives of the Study

- What are the primary objectives of the study?
- What are the secondary objectives of the study?

Study Design

- What is the study design?
- Why has this particular study design been chosen?

Study Population

- What is the study population?
- What are the inclusion criteria?
- What are the exclusion criteria?
- If it is anticipated that certain individuals may be excluded from participation after enrollment, what criteria are being used for discontinuation?

Vulnerable Participants

- Are vulnerable participants to be included in the study? If so, which specific groups of vulnerable participants?

- What is the justification for including vulnerable participants?
- Can the study be completed without including vulnerable participants?

The Specific Nature of the Drug(s) or Device(s) Being Studied

- What drug or drugs are to be used in the study?
- If more than one drug is used, have the drugs been used in combination before?
- Will a placebo be used in any part of the study? If so, what is the justification for its use?
- Have the safety and efficacy of the drug been established?
- Is the device being described in enough detail that the participant can substantially understand it?
- What does the device look like (e.g., size and shape) and how does it work?
- Where in the body will the device be inserted, and how?
- What will be required of the participant to have the device inserted (e.g., anesthesia)?
- What will be considered a positive outcome if the device is working properly?
- Will the device need to be serviced or replaced? If so, what will happen in the process?
- How will the participant know if the device is not working?
- What will be required of the participant if he or she suspects the device is malfunctioning?

The Risks of the Drug(s) or Device(s) Being Studied

- What are the foreseeable risks of the study drug(s) or device(s)?
- What is the chance that a risk will materialize?
- What may happen if the study device malfunctions?
- If a device malfunctions, what are the chances that the participant will survive, be permanently injured? Who will pay for hospital bills, clinical follow-up, physician fees? What compensation will the participant receive if he or she sustains irreversible injury?

The Dosage and Administration of Drugs Used in the Study

- What is the dosage and administration of drugs during each period of the study?
- What is the rationale for the selection and timing of doses?
- What type of blinding is used: single blind (the participant does not know whether he or she is receiving a study drug or a standard drug or placebo)

or double blind (neither the participant nor the principal investigator and research team knows whether the participant is receiving a study drug or a standard drug or placebo)?

- Will participants be allowed to take concomitant therapy for the treatment of other diseases while in the study? If so, why? If not, why not?
- How will compliance with treatment be ensured? What will happen if a participant does not comply with the treatment?

Experimental (Research) Interventions Used in the Study versus Clinical Care Interventions Occurring during the Course of the Study

- What interventions are being discussed in the scientific protocol?
- Are the interventions necessary for the clinical care of the study participant, and would they be conducted in the care of the individual even if he or she were not enrolled in a study? Or, are the interventions being done solely for research purposes and would not have been conducted on the individual if he or she were not part of a study?
- Are the interventions being done solely for the purposes of participant safety?
- How invasive are the interventions, and what are their risks?
- Could the same results be achieved with less-risky interventions? Why or why not? What are the trade-offs?

Endpoints/Outcomes of the Study

- What are the primary measures of efficacy being used?
- What are the secondary measures of efficacy being used?

Monitoring Data and Safety

- What are the risks of the study?
- How will the study participants' safety be monitored?
- How will study participants and their primary care or specialty physicians be alerted if an adverse outcome occurs?
- Who will be in charge of the management of each study participant's adverse outcomes and its follow-up and treatment to resolution?
- What are the criteria for determining when the study is too risky to continue and must be terminated?
- Who will follow individuals with adverse outcomes after participation in the study?
- Who will monitor data and how?

Statistical Methods

- How many participants are to be enrolled in the study?
- Is this sample size justifiable, given the risk to humans?
- What analyses are used to calculate success or failure of the study?

References

- What sources from the peer-reviewed medical literature and unpublished research findings are being used to support the reasons why this study should be conducted on humans and to document the chance of success in the study and the determination of both the nature and chance of risks that are cited by the principal investigator?
- What sources are used to support how all adverse outcomes are to be identified as early as possible and managed to the best extent possible?

Scientific Methods

As used in IRB review, the term *scientific methods* refers to the procedures the principal investigator is using in the study. Scientific methods need to be looked at from four angles: (1) the principal investigator's choice of methods, (2) how the methods translate into daily activities for the study participant, (3) who on the study team will interact with the participant regarding the requirements of the study at each stage during the study, and (4) who on the study team will interact with the participant after the study results are analyzed and interpreted. Here are some questions to bear in mind:

Box 5.2. Questions about the Scientific Methods

- How many participants are being enrolled in the study?
- Are appropriate individuals being included in the study?
- Are inappropriate individuals being excluded from the study?
- Are appropriate laboratory tests and studies being done during the enrollment phase to exclude individuals who are at risk (e.g., from prior exposure to a substance or from existing damage to liver or kidneys)?
- Are appropriate monitoring methods and response criteria (response of participants to study intervention) specified, and do they ensure optimal safety in the care of participants?
- Are study participants' laboratory values (results of clinical diagnostic tests) and study values being checked promptly and acted on as soon as they are available?

- Are laboratory and study values being checked too frequently or not frequently enough? This decision must be coordinated by members of the research team who have expertise in the medical conditions being identified through this laboratory and study testing. (Too frequent blood tests and radiological studies may place too much of a risk burden on participants. Too infrequent blood tests and radiological studies may result in delayed detection of new abnormalities.)
- Who is checking the laboratory and study values, and who are the best people to be evaluating and acting on abnormal values as soon as they are detected?
- Who is communicating with the participant's physician about abnormalities detected?
- Which physician is dealing with the abnormalities and caring for the participant with abnormalities?

Science in general can be roughly categorized into two types: observational and interventional.

Observational Science

In observational science, the scientist observes and records observations over time for study, analysis, interpretation, and reinterpretation. In clinical medicine, the early descriptions of most disease processes are observational, in that physicians observe and record over time their observations of a patient or patients and then study, analyze, interpret, and reinterpret their observations.

Interventional Science

Interventional science is the type IRBs typically review. In general, a research study viewed as interventional science follows a sequence:

1. initial observation period lasting several days,
2. an intervention of some type,
3. further observation over time,
4. end of study, and
5. long-term follow-up.

Initial Observation Period. The individual is observed during a medical history, physical examination, laboratory tests, and studies to determine his or her condition when the person is considering participation. (These are often called "baseline observations.") In addition, the individual may be asked to complete questionnaires to assess baseline psychological states. The information is then interpreted, and a decision is made whether the individual fits inclusion or exclusion criteria.

Intervention. An example of an intervention is when a study volunteer who has been identified as having an elevated blood pressure and has never been placed on a prescription medicine to control blood pressure is randomly assigned to receive either the study drug or one of the best drugs already on the market for the control of blood pressure, the comparator drug for this particular study. The object of the trial is to see whether drug 1 or drug 2 does a better job of controlling blood pressure with fewer side effects and with fewer adverse outcomes.

Further Observation over Time. Participants in both arms of a study are observed over time and additional medical histories, physical examinations, laboratory tests, and studies occur and are recorded to measure the effect, if any, the intervention had. In the blood pressure drug trial, the volunteers are followed to see how their blood pressure is controlled and to see if any problems occur with electrolytes, kidney function, or other parameters being measured by the research team.

If a participant's blood pressure goes unacceptably high, the individual may need to be terminated from the study and started on alternative therapy for control of the blood pressure. Similar action will be required if the study volunteer's kidney function begins to be impaired.

The principal investigator should at least recommend a treatment in the above situations. A recommendation may seem insufficient, since the principal investigator is responsible for seeing that adverse outcomes are handled with minimal damage to participants. The principal investigator must notify the participant's physician at once if the research team will not make a recommendation concerning the care of the participant in regard to the adverse outcome. There must be experts on the research team who are highly qualified, not only to conduct the study, but also to act quickly in the participant's best interest in the case of abnormal study results.

In a multisite study, the principal investigator must also report adverse outcomes to the study sponsor so that other sites where the study is being conducted can be warned. Computer interconnection of sites is necessary to make such cross-checks effective. The IRB needs to know how quickly such notifications are happening in all research studies their medical institution is involved with. There is no excuse for delay in notification, as it is of utmost importance to notify all participants of new adverse outcomes, to make certain the volunteers wish to continue in the study after hearing about the adverse outcomes.

End of Study. At the end of the study, the data are analyzed, statistically examined, and interpreted, and tentative conclusions are drawn as to whether and how the participants benefited or were harmed.

Long-term Follow-up. Some studies do not simply end. For example, the long-term effects of a new drug cannot be determined in a one-year study period. A drug may cause bone marrow damage, for instance, that does not show up until years after the study technically ended. Therefore, IRBs must be alert to the possibility of adverse outcomes after the study has ended.

The Randomized Controlled Trial

One of the key study forms that provides high-quality scientific evidence is the randomized controlled trial, in which participants are assigned randomly (for example, by the flip of a coin, by the roll of a die, or by a computerized randomization program) to one of two groups, one of which is the control group, against which the other is compared. A randomized controlled trial can compare new drug therapy with standard drug therapy, a new drug with a dietary or herbal supplement, new surgical techniques with standard surgical techniques, or dietary approaches with each other (for example, a low-carbohydrate–high-protein diet versus a high-carbohydrate–low-protein diet).

Some studies compare the substance being studied to a placebo (an inert substance without any intrinsic biological activity). The issue of when a "placebo only" group is appropriate in scientific research is under debate. It can be argued that it is necessary for a placebo to be present in the testing of a new drug because it is essential to demonstrate the efficacy of the drug, that is, that the drug has a benefit and that the benefit does not appear in the group of volunteers randomized to take the placebo. Placebo studies are required by the FDA for certain kinds of substances. However, physicians and other health care providers also need information on how new drugs compare to other drugs on the market for the same condition. Head-to-head drug comparison trials are not required for FDA approval, so a product manufacturer does not have an incentive to test its drug against others on the market for the same medical condition.

Each kind of study offers a different type of information. It is important to know the side-effects and benefits of a drug, and testing it against a placebo is good for that purpose. It is also important to understand how drugs contrast with each other in their risks and benefits to particular populations. When a new prescription medicine becomes available, the physician and the patient do not have much firm scientific basis for comparison with the best drug already on the market. At least with the latter there has been longer use of the drug in the general population, so there is a better chance that all of the risks have been identified. The principal investigator must clearly document the justifications for the use of a placebo. The IRB must decide if these justifications are compelling enough to warrant the use of a placebo and must discuss what criteria it will use to determine when a reason is compelling. Which kind of study to do needs to be addressed as a key research issue in all medical centers. The question is which drugs should get placebo studies and which need a head-to-head comparison.

It is not the case that all study drugs that exhibit a benefit over placebo during clinical trials will continue to show a benefit when restudied against other drugs. Perhaps the study drug showed only a modest benefit over placebo when used in highly selected participants and then fails to show even a modest benefit when tested in patients in general. Clinical trials against a placebo may not give clinicians enough information that they can confidently recommend the use of that drug to their patients.

The Selection of Individuals for Study

IRB members must ask themselves, Why is this study being proposed in human participants? The principal investigator must clarify why this research is important enough to be carried out with human subjects. He or she must fully reveal the purpose of the study; generalizations can be considered coercive in that the investigator could be seen as directing the participant's attention to the achievement of some desirable goal and not giving sufficient emphasis to the reality of the chance that an adverse outcome may occur. IRB members must ask themselves, Do I understand why an individual should consider participating in this study? The principal investigator must be able to give a compelling answer.

After the use of human subjects has been justified, the IRB examines the selection process laid out in the scientific protocol. The selection of individuals to participate in a study typically occurs in two stages: the decision of who is to be studied and why, then the choice of inclusion and exclusion criteria. The IRB looks at these two stages from both scientific and ethical perspectives.

Who Is to Be Studied? Scientific issues in general should be understood across both genders and all races. For example, only recently have scientists begun to rigorously study heart disease in women instead of assuming that heart disease in women is the same as heart disease in men. Likewise, osteoporosis, which is more common in women, also needs to be understood in men. IRB members must ensure that the principal investigator is including participants of both genders and all races or can justify why not.

Inclusion and Exclusion Criteria. Inclusion and exclusion criteria, the bases upon which eligibility for a study are decided exist for two main reasons: to ensure that individuals at moderate or high risk are not included in the study (except under special circumstances) and to yield results that are interpretable. To determine if the inclusion and exclusion criteria adequately define the appropriate study group, the IRB needs to question the principal investigator and local and national experts and, if necessary, conduct a systematic search of the peer-reviewed medical and scientific literature. In reviewing the inclusion and exclusion criteria, the IRB must pay particular attention to issues of justice and participant safety.

In the best studies on newly developed prescription drugs, not all types of participant have been included. Those typically excluded are individuals who have medical problems involving many organ systems and individuals who are taking many medications. Including these individuals in the study of a drug would make it difficult to determine whether any benefit or harm that occurred to a participant was due to the new drug. If an adverse event did occur to a participant with many medical problems, how would one tell whether the drug being studied had caused the worsening of the person's condition or whether this was a natural worsening? From a scientific standpoint, the ideal participant has no preexisting problem with functions of the liver or kidneys (the organs most involved in metabolizing drugs) and no disease other than the one being studied. Then, if a benefit accrues to the participant, it is more likely ascribable to the new drug because there are fewer competing reasons for the benefit to have occurred. If an adverse outcome occurs (e.g., liver or kidney function problems develop), it is more likely ascribable to the new drug. Yet if participants have been selected because they have only one disease (the disease of interest in the study), it should be realized that, as a group, they will be much less complex medically than the patients who will be receiving the drug once the FDA approves it. Thus, many questions will be left unanswered regarding how the drug will perform when used by more complex patients.

There are many other questions to consider in the choice of inclusion and exclusion criteria. Will individuals who are already carrying a burden of risk in their lives be included in the study? Will the appropriate blood tests and imaging tests be conducted before individuals enter the study, to ensure that they will not bear unnecessary risk or additional risk? If the research hypothesis involves exposing a participant to a toxic substance, will the appropriate medical history be taken and the appropriate laboratory testing be done to ensure that: the individual has not previously been exposed to that substance; the individual does not have an underlying medical condition or problem with the organ(s) needed to metabolize that substance; the individual is assured that a purposeful exposure to the toxicant is ethically justifiable and the assurance is scientifically valid?

Vulnerable Participants. Additional issues regarding inclusion and exclusion criteria involve vulnerable participants. The *Code of Federal Regulations* specifies the following individuals as likely to be vulnerable to coercion or undue influence: children, prisoners, pregnant women, mentally disabled persons, or economically or educationally disadvantaged persons.[2] Any study involving vulnerable participants must describe the rationale and justification for including members of a vulnerable population, and the principal investigator must specify the safety measures that will be employed to protect them. The class of vulnerable participants includes individuals whose decision-making capacity is compromised, such as those who have recently sustained a stroke or who have an ongoing condition that could impair their capacity to make a decision on their own behalf, like bipolar disease.

For vulnerable participants, the *Code of Federal Regulations* specifies: "additional safeguards have to be included in the study to protect the rights and welfare of these subjects."[3] Additional safeguards might include, for example, having a participant's spouse or next-of-kin be involved as a surrogate decision maker for an individual who has just had a stroke when consideration is being given to using a study drug intended to help persons with stroke. In stroke studies, delays can result in extensive and severe damage to the patient, so waiting for the patient to recover his or her full decision-making capacity might eliminate any potential benefit that might be received.

Surrogacy is a complicated legal matter. If the IRB does not have an attorney among its members, it must have ready access to an attorney with general expertise in human participants research and particular expertise in the federal and state laws pertaining to surrogacy that may need to be clarified in relation to competency and decision-making capacity. The IRB needs to determine whom it considers to be appropriate surrogate decision makers. This issue of surrogacy should be addressed by the entire medical institution with appropriate legal counsel, to ensure that all arguments are considered and appropriate approaches are mutually agreed on.

The *Code of Federal Regulations* recognizes both educationally and socially disadvantaged individuals as vulnerable participants.[4] IRB members must ensure that the participant is presented with materials at an educational level appropriate to the person, with the aid of translators for individuals who do not speak English as a native language or who are uncomfortable with English, and with ethical consultation regarding any issue for which there is a question involving how a particular vulnerable participant can be best protected.

The IRB must be quick to note problems in the inclusion and exclusion criteria and to give feedback to principal investigators and study sponsors. If the sponsor does not want to use appropriate inclusion and exclusion criteria, the IRB must not approve the study.

Sample Size and Analysis of Results. The best science uses the best statistics. Every IRB member needs to know or be willing to learn some basic statistics and its uses in research studies on humans. Clear thought about statistics should have occurred before the scientific protocol was ever brought to an IRB for review. Statistics should be considered during the planning phases of a research study when determining how many participants need to be studied to achieve statistically significant results. The IRB member with statistical expertise may be able to initiate a discussion of the statistical power of a study that may lead to a satisfactory IRB judgment about the study.

An IRB may need to consult statistical experts regarding the number of participants that should be studied to answer a particular scientific question. In a study that everyone agrees is important in developing new scientific knowledge, one does not want so few participants that statistical determinations cannot be made on the data. Nor does one want to enroll too many participants, particularly when there is risk associated with the study and when statistical experts may consider a smaller number to be sufficient. The key is to sample just enough participants to ensure statistical confidence in the results.

The Study Drug

What Kind of Drug Is Being Used in the Study? Drugs can be classified as:

- unapproved drug
  —not approved by the FDA for any indication
  —safety in humans not established
  —preliminary safety established but efficacy not established
- prescription drug approved by the FDA and being studied for an FDA-approved indication
- prescription drug approved by the FDA but not for the indication being studied
- nonprescription (over-the-counter) drug approved by the FDA
- herbal or mineral supplement (over-the-counter) (FDA can investigate complaints about adverse outcomes but does not consider for approval)

Does the Use of the Study Drug Require the Completion of a Special Form? If use of the study drug requires the completion of a special form (e.g., FDA form 1571 Investigational New Drug Application), has the principal investigator or study sponsor submitted the form? Has the FDA approved the drug for use in this study?

What Is the Intent behind the Study with respect to the Drug? The IRB must probe the intent of the study to determine whether the principal investigator or study sponsor has any economic motivation (e.g., getting a drug onto the hospital's formulary).

What Are the Risks of the Study?

The IRB must understand what the risks of the research study are. The list of questions in Box 5.3 will help IRB members analyze the potential for risks. The scientific protocol must spell out in clear and precise terms the adverse outcomes that may befall a study participant. The list of risks is based on a systematic search of the peer-reviewed medical literature and discussions with colleagues and experts. Death, stroke, headache, nausea, and vomiting are just a few examples of medical risks. Each risk should be specified, along with the best numerical estimate of the chance of occurrence.

Box 5.3. Questions about the Risks to the Participants

- What are the most severe adverse outcomes that could occur to a participant in this study?
- What are the common adverse outcomes?
- What are the worst states that could result from an adverse outcome?
- What monitoring criteria and interventions (including invasive medical interventions) may be necessary to gather the monitoring data? (For example, will a cardiac catheter have to be inserted under sterile technique with anesthesia to obtain data on pressure in the heart?)
- In the case of a study drug:
  —What are the risks of the drug?
  —What are the risks and projected benefits of the drug as compared to the risks and benefits of similar drugs and other drugs for the same condition now marketed?
  —What are the risks from drug washout (when a participant stops taking a current medication, with no new medication being added, before starting the study drug)?
  —What are the risks of drug-drug interactions?

- In the case of a medical device:
  —What are the risks of the device? (See Box 5.4.)
  —What are the risks and projected benefits of this device as compared to the risks and benefits of similar devices now marketed?

  The IRB needs to understand what the FDA has said about the device so far in its evaluation of the device.

  The IRB needs to understand how—that is, using what scientific principles—the device's manufacturer made its designation of risk and how those principles compare to other scientific and ethical principles used to identify and measure risk.

  —What is the device's estimated rate of failure?

  On what data was the estimate based?

  How accurate is the estimate?

  —What are the risks of the surgical procedure required to implant the device into a human being and the risks that exist while the device remains implants?
  —What will happen to the participant if the device fails?

Although some risks may be easily understood, others will need clarification and translation into language a participant can understand. For example, therapy with a drug used to manage atrial fibrillation may result in corneal microdeposits. On seeing such a statement, the IRB might ask the following questions, anticipating what may be of interest to the individuals being recruited as study participants: Why do microdeposits occur in the eye while the participant is taking a heart drug? What are the microdeposits made of? Is there any chance of permanent damage to the cornea? If so, how much chance? If the damage is not permanent, how long will the deposits last, and can anything be done to eliminate them? How will they affect the participant's quality of life?

Questions that are important in the consideration of risk related to an adverse outcome are:

- What foreseeable adverse outcomes may occur in the study?
- Does the study involve risk to only the individual participant, or does it involve risk to past, present, and future generations (as in genetic studies for which specimens are uniquely identified)?
- What is the pathophysiology of the adverse outcome?
- How will the adverse outcome alter a participant's life on a daily basis, and how will the participant feel different from before?

- Are there any particular types of participant who will be affected more severely if the adverse outcome occurs?
- Is there any management or therapy for the adverse outcome, and what is the chance that a particular individual who sustained the adverse outcome will benefit from such management or therapy?
- What was the prognosis of the participant before sustaining the adverse outcome, and how is the prognosis different after the adverse outcome?
- Are there ongoing research projects to better understand how to prevent, manage, and treat the adverse outcome in general and in specific high-risk groups or populations?

Risks Related to the Study Drug versus Risks Inherent in the Study. Even in a well-designed study, there may be risks beyond the adverse outcomes caused by the drug or device. There may be risks caused by the participant's stopping previous medication(s) and allowing them to be washed out of the body before starting the new drug (see just below).

How Safe Is the Setting for the Drug Washout? Typically, the safest setting for a drug washout is in a hospital, where the individual can be monitored. If the drug washout will not occur in a hospital, is the IRB sure that it can occur safely for an outpatient? The IRB should pursue this point with the principal investigator and search the literature to determine if and when a washout of the drugs these persons have been taking can occur safely on an outpatient basis.

Determining How Risky a Study Drug or Device Is. All IRB members must examine the scientific protocol to determine how risky the study drug or device is. As discussed in Chapter 3, risk has at least these key components:

- the nature of the adverse outcome that can occur
- the chance that the adverse outcome will occur
- the severity (or range of severity) of the adverse outcome that can occur
- the magnitude of the risk of an adverse outcome
- the participant's chance of recovering from the adverse outcome
- the estimate of whether the adverse outcome will be reversible or irreversible
- the need to monitor for any and all adverse outcomes with the goal of timely detection
- the alternative interventions (diagnostic or therapeutic) that an individual can obtain in routine clinical care without involving himself or herself in the study.

All *severe* adverse outcomes that can be envisioned should be stated, even if there is a low chance (probability) that they will occur. Severe adverse outcomes include, but are not limited to:

- death;
- stroke affecting the ability to move, speak, or think;
- scarring and other skin reactions that are permanent and irreversible;
- reactions requiring hospitalization in a cardiac care unit or an intensive care unit;
- adverse outcomes requiring intubation and ventilatory support;
- kidney failure, liver failure, diabetes, blindness, deafness, infertility, and birth defects.

It is difficult to define an order of severity for adverse outcomes, because views differ on what is more severe.

The IRB member cannot assume that a principal investigator or study sponsor is disclosing all known risks of the study drug. The completeness of disclosure of risk has to be verified by the IRB, consulting, as needed, local, regional, national, or international experts. In my opinion, all IRBs considering research on drugs or dietary or herbal supplements need to consult a clinical pharmacist. Ideally, they should have a clinical pharmacist on the IRB.

In the case of study devices, there are different kinds of risk.

Box 5.4. The Risks of Potential Harm from an Implanted Medical Device

- That the device will fail while implanted in the participant
- That the device will move around in the body if unseated from its original point of attachment
- That the device will produce inaccurate readings
- That, based on inaccurate readings, the device will perform incorrectly
- That the device will affect a medical diagnosis or treatment (e.g., if a participant with an implanted device goes to an emergency room whose physicians have no experience with the device)
- That the power supply for the device will be interrupted or shut off

Any of the events listed has the potential for causing harm to a participant.

Systematic Searches of the Medical and Scientific Literatures

The scientific protocol must provide documentation regarding the studies that the principal investigator and study sponsor used to support the scientific hypothesis, to select the scientific methods, to make the assertion that this study is important enough to involve human participants, and to document the risks stated. Interested parties can check the references, and the IRB should make certain that they are from current, peer-reviewed medical literature. Consulting these references will provide a wealth of data that may be useful to IRB members in their deliberations.

The necessity of a literature search is illustrated in the example of a study drug that has already been studied to a certain extent and reported on in the peer-reviewed medical literature. The risk section of the scientific protocol reports only the risks mentioned on the product's label. This is inadequate. A long time may elapse between drug approval and approval of a product label, and even when a label is approved, it may quickly be behind the research that is published related to the drug. The IRB can conduct its own search of the literature, looking for adverse outcomes associated with the drug, the nature of the adverse outcome, its chance of occurrence, populations at special risk, etc.

Each IRB member must learn how to search the peer-reviewed medical literature. When performing such a search, IRB members must understand the risk profile of each agent (drug or device) or intervention that is part of the research study, to determine whether the disclosures regarding the risks are all inclusive and will be clear and understandable to participants.

Systematic searches of the medical and scientific literatures can include searches for:

- adverse events, adverse outcomes,
- precautions,
- contraindications,
- monitoring (e.g., how frequently laboratory tests should be done to assess liver and kidney function), and
- drug interactions (with other drugs, with herbal or mineral supplements).

These types of information affect the inclusion and exclusion criteria and estimates of risk to the participant.

New IRB members should not consider searching the medical literature a daunting task. The institution's research service or the IRB, with the help of the institution's librarian(s), may hold a training session demonstrating how to conduct a search. Searches are sometimes slowed by the unavailability of a journal in the institution's library. Librarians can help the IRB member obtain the needed article or journal issue.

The results of a basic search should be reviewed with local experts, for instance, to extend the list of known risks. Clinical experts, by attending medical society meetings that report and discuss recent research, may have learned of risks before they appear in the medical literature. The IRB can then present the result of the basic literature search to the principal investigator, who will complete the review and add references to the proposal if needed.

Baseline Laboratory and Study Measurements of Participants

All studies involving drugs and devices require that study subjects have a baseline medical history, physical examination, set of laboratory measurements, and set of study measurements. These baseline studies are needed principally to assess the individual's physical and mental states at the time of entry into the study and to identify any impairment of function in the individual's bodily organs. (For example, because most drugs are metabolized and eliminated by the liver and kidneys, respectively, if the individual has an abnormality of the liver and/or kidneys, that condition may be worsened by the study drug.) The measurements will be repeated during the study, to determine if any new abnormalities have developed. The IRB may need to ask at what intervals these repeated measurements will be taken and whether they will be—or are being—done in such a manner that the participants are properly treated if problems appear.

The Participant's Involvement

A person's experience as a research subject can be divided into phases.

1. Pre-enrollment evaluation
2. Enrollment
3. Initial participation, as study gets under way
4. Continuing participation, as study progresses
5. Closing participation, as study draws to a close
6. Being informed of outcomes of the study after data have been analyzed, interpreted, and published in the peer-reviewed medical literature
7. Participation beyond the study (e.g., regarding long-term effects of the intervention on organs and organ systems or the long-term effects of investigational devices left in the participant's body)

The scientific protocol must clearly state how long the participant will be enrolled in the study. It must answer the following questions:

- How long will an individual participant be enrolled in the study?
- How many times or how frequently will the participant have to come to the medical center or study facility?
- How long will the participant spend during each visit to the medical center?
- What will the participant be doing in relation to the study while at the medical center?
- What will happen to the participant once the study is completed or stopped? (For example, will the participant continue to receive the study medication if the study is successful? If so, for how long and at what cost? If not, why not, and what will happen to the participant?)

How Many Times Will the Participant Have to Come to the Study Facility?

The IRB member needs to understand how much of the participant's time will be involved in the study. This requires focusing on time points (visits to the institution for study-related activities), time intervals (length between visits and length of each visit), and total duration of the study, including any study-related activities that may occur after the study. The issue of duration also includes whether the principal investigator or research team will contact the participant regarding incomplete or missing data or to see how the participant is doing physically and mentally after the study has been completed.

What Will the Participant Be Doing while at the Study Facility?

The IRB needs to be able to tell exactly what the participant will be doing at each visit. For example, how many questionnaires will the participant be asked to complete at each visit, will emotionally draining questions be asked, and what will be done to support the participant if he or she becomes upset while completing a questionnaire? Will blood be drawn, and, if so, how much will be drawn at each visit? Will there be x-ray studies, and, if so, to what cumulative radiation doses will the participant be exposed?

What Will Happen to the Participant Once the Study Is Completed or Stopped?

Multiple issues arise at the end of a study. In a new drug study, for example, if the participant has done well while taking a new drug, will he or she be allowed to continue taking the drug? If so, what will be the cost to the participant? If not, what alternatives will be offered to the participant whose medical condition might not be controllable with drugs currently on the market? If the participant does not have a primary care provider, who will follow the participant until a suitable treatment is found?

If the principal investigator and study sponsor cannot or will not provide the study drug after the study ends, the participant needs to know and understand the consequences of this *before* enrolling in the study, and the IRB must make sure that the informed consent form and session are fully informative in this regard. If the principal investigator and study sponsor elect to enroll a participant who does not have a health care provider, then the principal investigator and sponsor should be held accountable for finding a health care provider willing to care for the participant. If the scientific protocol does not spell out the provisions for post-study care of participants, then the IRB should not approve the study, because the IRB's task of optimally protecting participants extends beyond the close of the study to successful transfer of the participant with a well-managed condition back into the community.

Other Issues Involved in the Review of a Scientific Protocol

Specification of Length of Study

Box 5.5. The IRB's Concerns about the Length of Studies

- What constraints on time length should a study have?
- Can a principal investigator or study sponsor justifiably specify a completion date as indeterminate?
- Certain studies seem to go on forever. Is this legitimate?
- Can a completion date of a study be defined in terms of the number of participants to complete the study (e.g., after 500 volunteers have finished the study)?
- Can a study plan to follow a participant until death?

Each research study must have a start date and a completion date. The IRB member will need specific information about the start and completion dates and what intervening events might occur to alter these dates. If the principal investigator wishes, he or she may later request to extend the project (e.g., because of difficulties in achieving the desired enrollment in the study).

Who Will Be Recruiting Participants for the Study?

The protocol should specify who will be contacting individuals and asking them to consider participating in the study (the principal investigator, a research nurse?). Will these same people be conducting the informed consent process? What is their training, in general and in recruiting participants and conducting informed consent sessions? Who is responsible for overseeing their performance? Can the principal investigator provide authoritative evidence of the competency of these personnel? Are the personnel sufficiently familiar with the study to be able to answer any question the potential participant may have? If not, who will answer the participant's questions?

Auxiliary Personnel

The scientific protocol should include clear and complete instructions for auxiliary services whose personnel are not part of the research team but who are essential to the appropriate and safe conduct of the research study. Such personnel include laboratory and diagnostic services, supply services (which may be responsible for the distribution of devices or diagnostic and interventional instruments used in the study), and pharmacy services or whatever institutional entity will be responsible for the dispensing of the study drug.

Lack of Clear and Precise Definitions

Although the IRB must ask the principal investigator to provide clear and precise definitions in all aspects of the science, there are many instances in which terms are not clear. The onus of presenting all definitions and translations in lay language falls on the principal investigator and research team. The principal investigator may have to work extensively with colleagues to develop clear and precise definitions of all terms used in the scientific protocol so that they are understandable by nonscientists. Two concepts are of particular interest to IRBs.

Reasonably Foreseeable Risk. Federal regulations specify that informed consent forms disclose reasonably foreseeable risks,[5] yet, the *Code of Federal Regulations* does not provide a clear definition—that can be immediately put into practice—of what a reasonably foreseeable risk is. Basically, reasonably foreseeable risks are those risks that are known to and predicted by experts. The principal investigator should be required to be an expert, to have a relevant expert on the research team, or to consult relevant experts.

A scientific protocol should disclose all risks. The IRB may need to consult local experts or to search the peer-reviewed medical literature to ascertain all the risks of a drug or medical device. If a drug is new, what are the risks associated with the class of drugs which the drug most resembles? Even FDA-approved drugs have associated risks.

Reasonably foreseeable risk is extremely important and can be tricky to estimate. The FDA usually studies drugs in 4,000 to 6,000 subjects before approval.

Once the FDA's approval process is complete and the drug is marketed in the general population, additional adverse outcomes are likely to be identified and unlikely outcomes to occur. The chance of dying from a drug may be 1 in 20,000 patients. Until the drug has been studied in 20,000 patients or more, that one death may not occur, and it may happen as 5/100,000 or 50/1,000,000. After a drug is marketed, new adverse outcomes become topics of discussion in the peer-reviewed medical literature. When assessing research on such drugs, IRBs need to seek out the very latest and fullest data on the substance.

Unknown Effects. Does the principal investigator specify that there is the potential for unknown effects, especially unexpected adverse outcomes, to materialize during the study that may be related to the study drug or device? The IRB member must ask if the effects are truly unknown or are expected based on similar earlier research study results. That is, in addition to the vague notion of reasonably foreseeable risk, the fact is that unexpected adverse outcomes may occur during the study.

The problem with unknown effects is that what is unknown may be relative to the quality of the search of the medical and scientific literatures on previous studies with similar agents. The principal investigator must thoroughly search the peer-reviewed literature. The IRB also must conduct its own search of the literature.

The Principal Investigator May Not Have Reviewed the Complete Submission

The IRB must insist that the principal investigator take full responsibility for the submission of the scientific protocol and informed consent form. It is unacceptable if the principal investigator does not completely understand the research that he or she is proposing to do. The principal investigator is responsible for understanding not only every detail of the study and why it is being proposed but also his or her role in the protection of all participants in the study.

Multisite projects can get complicated. The scientific protocol and informed consent form are written by the principal investigator who instigates the study or a sponsor who recruited him or her. The research proposal must be considered in its entirety within each site institution. Each site has a principal investigator, and it is the local principal investigator's responsibility to present the proposal to the local IRB and to contact the initiating principal investigator to obtain answers to all questions that the IRB raises about the scientific protocol and informed consent form.

Approval by One IRB Does Not Imply Approval by Another

The fact that one institution's IRB has approved a research study does not guarantee approval by another IRB. Similarly, the fact that a scientific protocol and informed consent form come from a different site does not mean that they have been approved by that other site's IRB. Federal regulations mandate local review of research. Each IRB must carefully review and evaluate each research proposal submitted to it. A scientific protocol may be acceptable to one university or medical center but not to another, as may the suitability of the principal investigator's credentials in his or her role in overseeing the project and protecting study participants. Principal investigators must be aware that a research study involving human participants approved by one institution's IRB may be rejected by another IRB.

Although multiple reviews are laborious for the principal investigator, a study can benefit from another IRB's review of the scientific protocol and informed consent form. Fresh eyes may offer new insights and suggestions for revisions and changes. Indeed, the project may benefit from expertise that is present at one institution but not at another.

Links between the Scientific Protocol and the Informed Consent Form

There must be a one-to-one correspondence between the issues in the scientific protocol and a clearly translated description of the same points in the informed consent form. This is not to say that each point in the scientific protocol must be stated in the informed consent form; rather, it means that a principal investigator cannot, in an informed consent form, contradict or misstate what will happen in the research study as described in the scientific protocol. In addition, the principal investigator must provide in the scientific protocol all the necessary information, not a subset of the information. The principal investigator should never hide from the IRB any information related to the judgment of a study proposal. The goal in communication with the IRB should be openness and honesty at all times.

Challenges for the Nonscientist

The review of a scientific protocol may be particularly challenging if the IRB member is a nonscientist. Often an IRB member does not understand an issue in its entirety but hesitates to raise an issue because it seems not to be a problem for others. Any time he or she does not understand something, an IRB member has an obligation to speak up during the full board meeting. One person's question may inspire questions in others' minds. Many times, the principal investigator is asked to come to a meeting to discuss an issue, or experts inside and outside of the institution are asked for opinions. This process often leads to discovery of issues and solutions that would otherwise go unnoticed and unresolved.

The Scientific Protocol Is Probably Written in the Language of Scientists

If the scientific protocol is written in language that is heavily scientific, it can be a problem for the nonscientists on the IRB. Therefore, the IRB must insist that scientific protocols be written in language that nonscientists can understand. Otherwise, how will the nonscientist participate in IRB decision making related to the study? It is not enough to say that the concept can be explained to the nonscientists during the meeting. Rather, the principal investigator must make every effort in careful writing and translation of the scientific protocol into language that can be understood by nonscientists. After all, the science is going to have to be made understandable to potential participants in the informed consent form, so the clarification and translation will need to be done eventually. The onus of careful writing and clear translation in both documents is always on the principal investigator. Nonscientist IRB members should be quick to say when they do not understand the scientific protocol, because many of these issues will recur in the informed consent form.

Abbreviations

One of the most frustrating situations for a nonscientist on an IRB is to encounter abbreviations in the scientific protocol. Abbreviations are used commonly in clinical medicine and scientific research (for example, COPD for chronic obstructive pulmonary disease and GERD for gastroesophageal reflux disease). IRB members must ensure that all abbreviations are defined understandably. In writing the original study proposal and informed consent form, the principal investigator needs to consider the audience for the informed consent form and define abbreviations. However, the IRB member must be ready to help with the task of translation as well. Because of the IRB's expertise in this area, the IRB should consider developing a glossary of abbreviations to aid new and nonscientist IRB members in their work on the IRB.

Obfuscation

Scientific terminology can be more than just a barrier to understanding. The IRB should watch for any attempt in a scientific protocol to divert attention from underlying information by the use of general or categorical terms instead of the specifics in discussing adverse outcomes that could befall a study participant. This kind of problem may be difficult for a nonscientist to detect. For example, *blood dyscrasia* is both a complex medical term and a categorical term related to a general class of blood problems. The use of such a term could be an attempt to avoid revealing more detailed information, in this case whether and to what extent the drug in question causes any of the following specific types of blood problem:

- a decrease in platelets that may result in an increased tendency to bleed
- a decrease in red blood cells (that is, causing an anemia), making it more difficult to breathe and survive if the anemia is bad and persists
- a decrease in white blood cells that may result in an increased risk of infection.

The IRB member must be alert for presentations that use ambiguous concepts or compound terms that obscure what is going on, and he or she must demand clear translations of all concepts and terms. Again, the practice of speaking up when something is not clear will lead to a clarification of terms that will benefit the quality of the protocol and informed consent form and the comprehensiveness of the IRB review.

6 The Informed Consent Form

Federal regulations mandate that the informed consent form, read by potential study participants and signed by those who enroll, must disclose certain information that will help individuals decide whether or not to enroll in the study. Additional information should help the individual understand and appreciate the research and the legal nuances of agreeing to participate in research. The information is provided in the attempt to allow the individual to make a considered decision to participate. Still, the reasons people enroll in research studies vary by the individual.

Informed consent is both an event at a point in time and a process over time. An individual being recruited into a research study has an informed consent session with the principal investigator or his or her designee. A copy of the informed consent form is given to the potential participant at an informed consent session, to allow him or her to deliberate on the specifics of the study and to generate questions for future discussion with the principal investigator, research team, or representative of the IRB. The informed consent process continues throughout an individual's participation in a study and even afterward if new information is obtained after the person's part in the study has been completed. However, the informed consent form should be complete and not rely on assumptions about what the principal investigator may or may not tell a potential recruit in an informed consent session or subsequently.

There should be no attempt to coerce an individual to participate in a study. Some principal investigators or designees who are soliciting the consent at the informed consent session do engage in what could be described as convincing, logically arguing, or gently persuading. The best approach is to allow the potential recruits to include as many others in the discussion as they wish and then decide to enroll on their own.

The procedures used in obtaining informed consent should be designed to educate individuals in terms that they can understand. The informed consent form and its documentation (especially the explanations of the study's purpose, duration, experimental procedures, alternatives, risks, and benefits) must be written in lay language.

Often the informed consent form lacks key information because the principal investigator is afraid that if the individual deliberating about study participation is aware of this information (for example, the risks of the study) he or she will not enroll. The IRB must see that the informed consent form includes and substantially discusses all pertinent information, including all risks, and does so in an understandable way. The IRB is assessing the informed consent form along with the scientific protocol, which contains much information that will not have been included in the informed consent form.

In Appendix 2 at the back of this book is a check list for reviewing informed consent forms.

The Structure of the Informed Consent Form

A basic structure underlies the sections of an informed consent form. When examining the form, an IRB member may find the elements shown in Box 6.1.

Box 6.1. Elements of the Informed Consent Form

1. Principal investigator, study sponsor, and overall design of the study
2. Purpose of the study
3. Co-investigators
4. Procedures and involvement of the participant
5. Potential benefits, if any
6. Potential risks
7. Potential discomforts
8. Section for women, if there are particular risks to women, including risks to fetus during pregnancy
9. Section for men, if there are particular risks to men
10. Liability (who is going to pay for damages incurred by the participant during participation; what is and what is not going to be compensated)
11. Alternatives to participation in the study
12. Participant's health information (particularly involving confidentiality and authorization to disclose elements of the health information)
13. Participation and withdrawal
14. Costs and compensation to the study participant
15. Openness to participant's questions
16. Participant's rights
17. Biological specimens (if any will be involved)
18. Tissue banking (long-term storage of human tissue, with or without unique identifiers)
19. HIPAA-protected health information disclosures

After specifying the study title and study objective or hypothesis, the informed consent form should address the following areas.

Study Procedures. The purpose of this section is to translate the "Study Methods" section of the form into the precise procedures that participants will be involved in, for example, questionnaires and surveys to be completed (the form should estimate the time to complete), and blood drawing or tissue removal.

Study Risks. This section lays out in clear detail the risks the individual faces once he or she agrees to enter a research study. The concept of research risk is complex. Thus, the risk section should include reference to many aspects of adverse outcomes and their occurrence. Both the nature of possible adverse outcomes and the best numerical estimate available of the chance (probability) that each adverse outcome will occur should be included. This section must avoid minimizing any risk the participant may encounter and must include the fact that there may be additional risks that principal investigators and study sponsors will not be aware of until the study is undertaken.

The impact of research on the participant is measured in terms of the chance of dying and the chance of change in quality of life (for example, the chance of a stroke resulting in partial or complete paralysis, the chance of a stroke resulting in sensory impairment, or the chance of a stroke resulting in cognitive impairment). Some informed consent forms have a "Risks and Discomforts" section instead of labeling it "Risks". I prefer to see the risks separated from the discomforts, so that individuals being recruited into the study can clearly distinguish discomforts—that is, mildly adverse outcomes—from more severe possible outcomes of participation in the study.

Alternatives to Study Participation. This section clearly must state that enrollment in a research study is completely voluntary and that, should the individual not wish to enroll, all alternative standard treatments for the medical condition being studied available at the medical institution will still be open to him or her.

Confidentiality and Anonymity. It is not sufficient for the principal investigator simply to state that confidentiality and anonymity will be maintained; he or she must explicitly state what will be done at each point to give the best chance of maintaining confidentiality and anonymity. Confidentiality can best be maintained by keeping all data files and specimens in securely locked areas in securely locked offices in securely locked and monitored buildings. This section of the form should specify who (for example, FDA, study sponsor, or others) will be examining the participant's study records and what elements of the records they will review. Anonymity can best be maintained by never associating a unique identifier with any sample or questionnaire. But however well intentioned the investigator may be, even the most secure safeguard may be broken. This possibility will need to be admitted in the informed consent form, with a specification of what will be done if that happens.

Costs to the Study Participant. Some studies may involve direct financial costs to the participant. Any such costs (for example, copayment for drugs used to treat side effects of the study drug) must be listed and explained.

Liability Assumptions. This section states who, if anyone, will pay for physician and hospital costs if an adverse outcome occurs that requires hospitalization and who, if anyone, will provide compensation for long-term disability if an irreversible adverse outcome (such as a stroke) occurs.

Difference between Research Participation and Clinical Care. This section explains that participation in research is not clinical care and how the purpose of research (the development of generalizable knowledge for future patients) is different from that of clinical care (as defined in the *Code of Federal Regulations* and the Belmont Report).

Voluntary Nature of Participation. This section clarifies the point that all participation in research is optional.

Withdrawal from the Study. This section states that a participant can withdraw from the study at any time but must do so safely. The principal investigator must clearly discuss what would need to take place if the participant decided to withdraw from the study. For example, what happens if a participant who has had a device implanted wants to withdraw?

Other Rights of Research Participants. This section lists and explains whatever other rights the participant may have, for example, there may be relevant state laws that affect disclosures related to release of genetic data; these laws may grant a person additional rights beyond those protected at the federal level.

Contact Numbers and Names for Research Staff. A participant may contact the research staff with any questions concerning the scientific aspects of the study (see Box 6.2). This section gives the information and parameters for doing so.

Contact Number and Name of the IRB Chair. A participant may contact the IRB chair or designated member with any questions concerning the way the study is being conducted or with complaints regarding any aspect of research participation (see Box 6.2).

Drugs and Their Risks. If the drug under investigation is experimental, this section should state that and should specify whether or not it is approved by the FDA for the use being studied. A drug may be approved for one indication but unstudied for another. For example, a drug that is approved as an anti-inflammatory agent for use in the treatment of arthritis may be under study for decreasing the occurrence of polyps in the human intestine. Thus, this drug

Box 6.2. Areas of Potential Concern to Study Participants

- *Scientific questions about the research study.* A participant's questions about the study, its purpose, and the way the study is being conducted scientifically, at any point during the study, should be addressed to the principal investigator or a knowledgeable member of the research team.
- *Ethical questions about how the research study is being conducted.* Questions about how the research is being conducted from an ethical standpoint and how the research team is interacting with the participant should be addressed to the IRB chair or a designated IRB member.
- *Liability and compensation questions regarding research-related injury.* Questions about research-related injury can be addressed to the IRB chair, an IRB member, or the institution's attorney or staff.

would be considered experimental in this new use. There should be a subheading for each drug in the study and its risks and interactions with other prescribed medications and with over-the-counter medications and dietary and herbal supplements. If drugs or interventions can cause damage to a fetus, then the risks of pregnancy and risk to fetus must be included for each drug. There should be a statement regarding acceptable methods of birth control if pregnancy is possible.

General Issues concerning the Informed Consent Form

Readability

Individuals being recruited as study participants must be able to read and understand the entire informed consent form. Achieving such comprehensibility is not easy, because the individuals reading the document may come from a variety of backgrounds. They will have more or less experience with clinical care, with how scientific research in medicine is conducted, and in reading lengthy documents with many new concepts and frank statements of rights. The IRB must strive to educate investigators regarding readability at the level of the nonprofessional.

What to Say about Risk

Often, a study's scientific protocol contains a complete description of all risks of every drug involved in the study, with each risk assigned a best numerical estimate of its chance of occurring, but the informed consent form contains only a subset of the risks and uses vague probability terms (e.g., "some chance of" or "rare"). There are no simple guidelines to follow when there are many risks to disclose. A quick search of the peer-reviewed medical and pharmacy literatures will show that almost any drug carries many more risks than are listed in an informed consent form. Should all side effects be listed, or only the most common and the most serious but less common? Especially for the minor and moderate side effects, the IRB must decide which should be listed in the informed consent form and how much data needs to be conveyed about each; it should record its decisions, to see over time which work and which do not.

The notion of what is or is not a "foreseeable risk" is problematic. It is not the case that because there has not yet been a death connected with the new drug, death should not be listed on the informed consent form as a possibility. If the class of drugs to which it belongs, or a similar class, or the diseases or medical conditions have been associated with death, then study participants should be warned of the potential risk of dying. The details of other studies of the drug can be important; because a risk does not materialize in highly selected study participants does not mean that it will not occur in a broader population.

No Coercion

Coercion in any form has no place in the wording of an informed consent form. IRB members must learn to identify coercive language, with the goal of eliminating it.

Perhaps the most extreme example of coercion I have witnessed in an informed consent form was in a drug study sponsored by the drug's manufacturer, who wanted to guarantee participation to the end of the study. The consent form asserted that if a participant left the study early, the product manufacturer could hire a detective to track down the participant who failed to make him- or herself available to provide study-related information (data) to the study sponsor.

Another case of coercion would be if a principal investigator had a prominent politician endorse participation in a particular study. The order in which information is presented in an informed consent form can exhibit an element of coercion. Imagine that a drug manufacturer as a study sponsor places on the last page of the informed consent form a statement that the drug manufacturer will not pay for hospitalization for disabilities that occur during study participation if there is any question whether they are study related. Now imagine that the statement is placed on the first page. How many individuals would enroll in the study? An IRB member may rightly argue that any such statement should be removed and a full discussion of participants' rights placed prominently in the informed consent form.

Even the way in which sections of an informed consent form are printed can serve as a vehicle of coercion. For example, coercion may be present if boldface type and highlighting are used for a section on positive aspects of research participation whereas the risks section is printed in small, hard-to-read type.

Coercion may also be related to the payment of hospital costs for adverse outcomes and the payment of compensation for short- or long-term disability related to adverse outcomes. A study sponsor may not be willing to "offer" compensation for such injury and damage, but the participant always has the right of suit to recover compensation for damages. Therefore, an informed consent form cannot state that a study sponsor "will not pay" for hospitalization or damages related to research participation.

It may be coercive if there is no discussion of a certain topic in an informed consent form. If an informed consent form fails to mention who will pay for care if an adverse outcome occurs, this is persuasion by failure to mention an issue that might deter someone from participating.

Let us take another example of potentially coercive wording. Consider the statement, "This study has the approval of the medical institution's IRB." Some individuals may interpret the statement as an endorsement of the study as an activity in which the institution's patients should participate. This is not necessarily the case. A more typical interpretation and wording of the IRB's position is, "The scientific protocol and informed consent form have been reviewed and, after modification, have met the institution's criteria for a research study to be offered to individuals who meet the inclusion criteria." If a principal investigator or designee attempts to attribute more to this minimal statement, he or she is coercively manipulating the fact of IRB approval.

Stamping "IRB approval" at the bottom of the informed consent form could be interpreted as coercive because someone might feel that refusal to participate in the research would constitute a decision against an approving, official body of the institution. Also, an individual considering participation may misinterpret approval to mean safety and believe that the IRB is somehow ensuring that the study is safe. (In fact, it is primarily in the hands of the principal investigator and research team to minimize the occurrence of adverse outcomes.)

Consent Form Not for Recruitment

It is not proper to use an informed consent form as a device to recruit individuals into a study. Similarly, it is not proper to use an informed consent form as an advertisement. Conversely, the language of recruitment and advertising is not appropriate in the informed consent form. The latter should be a straightforward discussion of the nature of the study and all of its components.

Substantial Information and Substantial Explanation

The goal of the informed consent form is to provide an individual considering study participation with substantial information and substantial explanation of that information in an accurate and clear manner. The IRB must ensure that this goal is accomplished. "Substantial information" means, for example, that in a study in which the study drug may cause a syndrome, the informed consent form must provide substantial information about the nature of the syndrome and what that syndrome would mean to a study participant if it occurred over the short term or long term. The informed consent form should provide information on how the principal investigator and the research staff will look for the earliest occurrence of the syndrome in participants and what the temporary and permanent effects of the syndrome are.

Medical Terminology

The informed consent form must translate medical terms into lay language.

Legal Terminology

The issue of "legalese" (the abstruse technical vocabulary of the law) can arise with informed consent forms. Often, legalese is used when one party is attempting to limit its liability, such as when a medical institution's or study sponsor's attorneys try to limit the institution's or sponsor's liability if a severe adverse outcome occurs.

The major problem with liability is determining when an adverse outcome is in fact caused by a study drug or device. This process is difficult to describe in nonprofessional language; I have rarely seen this important issue mentioned, let alone clearly stated, in an informed consent form. The study sponsor and principal investigator do not want to scare away a potential study participant with lengthy discussions of a severe adverse outcome that has a low probability of occurring. Work needs to be done on how to translate into clear terms the language regarding legal liability and how any liability issues will be settled.

Basic Questions about the Informed Consent Form

Here are the basic questions that IRB members should ask about each key area of the informed consent form.

Principal Investigator, Study Sponsor, and Overall Design of the Study

- Does the informed consent form explicitly state who is the principal investigator in charge of conducting the study?

- Does the form explicitly state who is sponsoring (that is, paying for) the study?
- Does the form explicitly state what type of study the participant is getting involved in?
  —If studying a new drug, does the study compare a new drug to a standard drug, a new drug to a placebo, or a new drug to a standard drug and a placebo?
  —Is a new drug being studied for consideration of approval by the FDA? If so, which phase of the approval process is this study part of? (See Box 6.3.)

Box 6.3. The Phases of an Investigation of a Previously Untested Drug

The clinical investigation of a previously untested drug is generally divided into phases. Although in general the phases are conducted sequentially, they may overlap. The study phases leading to FDA approval are officially defined as follows (from www.fda.gov/cder/about/smallbiz/faq.htm#Definitions)

> Phase 1 includes the initial introduction of an investigational new drug into humans. These studies are usually conducted in healthy volunteer subjects. These studies are designed to determine the metabolic and pharmacological actions of the drug in humans, the side effects associated with increasing doses, and, if possible, to gain early evidence on effectiveness. Phase 1 studies also evaluate drug metabolism, structure-activity relationships, and the mechanism of action in humans. The total number of subjects included in Phase 1 studies is generally in the range of twenty to eighty.
>
> Phase 2 includes the early controlled clinical studies conducted to obtain some preliminary data on the effectiveness of the drug for a particular indication or indications in patients with the disease or condition. This phase of testing also helps determine the common short-term side effects and risks associated with the drug. Phase 2 studies usually involve several hundred people.
>
> Phase 3 studies are intended to gather the additional information about effectiveness and safety that is needed to evaluate the overall benefit-risk relationship of the drug. Phase 3 studies also provide an adequate basis for extrapolating the results to the general population and transmitting that information in the physician labeling. Phase 3 studies usually include several hundred to several thousand people.

Phases 1 through 3 of an investigation take the new drug to FDA approval. Phase 4 studies typically look at clinical effectiveness and safety of a drug after the manufacturer has met the requirements of approval and the drug has begun to be prescribed. Some Phase 4 studies may search for a newly approved drug's adverse outcomes, particularly severe adverse outcomes, and their rates of occurrence. In addition, the IRB may see post-approval Phase 3 studies designed to investigate new applications of approved drugs.

Co-investigators

- Does the informed consent form clearly identify the co-investigators and other research personnel the participant will be encountering in the study?

Purpose of the Study

- Does the informed consent form clearly state the purpose of the study and why humans are needed as participants?

Involvement of the Study Participant

- Does the informed consent form state the procedures the individual will undergo if he or she decides to participate in the study?
- Will the individual be interviewed by a study investigator or research team member to confirm that he or she meets study requirements?
- What will this interview include?
- Does the individual have the decision-making capacity to consider enrollment and does he or she understand what it means to voluntarily consent to study participation, what research is, what the risks of the study are, and the fact that the risks could materialize in her or her case?
- Will there be an emotional, physical, or psychological evaluation?
- Will there be a review of the individual's medical history?
- Will a physical examination be performed to ensure that the individual is in good health?
- Will blood be drawn? If so, what tests will be done, and why? How much blood will be drawn? When will the individual learn the results of the tests?
- Will other tests be done? If so, which ones and what are their risks? When will the individual learn the results?
- How many more visits will be required, over what time interval, and for approximately how long?
- Will questionnaires be required? If so, will the questions be personal and upsetting? What counseling will be available to individuals during or after the questionnaire sessions to help with their emotional state and support them regarding the questionnaire aspects of the study?

- If the study involves a placebo, does the consent form explain that a placebo represents nontreatment?
- If this is a new drug versus standard drug or a new drug versus placebo study, will the participant be randomly assigned to receive the new drug, a standard drug, or placebo? Is randomization adequately explained?

Potential Benefits, if Any

- Does the informed consent form clearly list and describe any potential benefits that may accrue to the participant? If there will be no benefits—apart from having contributed to the advancement of knowledge—does the informed consent form clearly state this? If the probability of receiving a benefit is very uncertain, is there a clear statement, "You may or may not personally benefit from participating in this study" followed by appropriate elaboration?

Risks

- Do the risks listed in the informed consent form, and their adverse outcomes, their natures, their chances of occurrence, and their impacts on a participant's survival and quality of life, match the risks listed and described in the scientific protocol? If so, go on to the next question. If not, the IRB member must ask the principal investigator why there is not full disclosure of risks in the consent form.
- Do the risks and their adverse outcomes, natures, and chances of occurrence match the IRB member's search of the peer-reviewed medical literature or searches of the literature available through relevant computer links or discussion with experts in the field? If not, the IRB member must ask the principal investigator to explain the differences. In particular, statements such as "There are risks that may not yet be identified with the study drug" should not appear if further risks are cited in the medical literature.

Potential Discomforts

- Are there potential discomforts not listed in the risk section that the participant should be aware?
- Is traveling or, conversely, not traveling a requirement of research participation? If so, this must be specified early as a condition of enrollment in the study. Have *all* restrictions that the study places on the individual been clearly thought through by the research team and explained in the informed consent form, so that there is a clear communication of expectations?

Section for Women, if There Are Particular Risks to Women or Pregnancies

- Is it known how this treatment could affect an unborn child? If so, the known risk(s) must be specified. If not known, the participant must be informed that "it is not known how this treatment could affect an unborn child."
- Does the form state as a requirement, "If you are sexually active and capable of becoming pregnant, you must use an effective method of birth control while participating in this study (birth control pills, IUD, diaphragm, condom with spermicide, abstinence, among others)?" Does the form also specify how long such methods must be employed?
- Is it known whether the study drug is distributed into breast milk? Are women who are or will be breastfeeding asked to refrain from breastfeeding while participating in the study and for a period of time afterward?

Section for Men, If There Are Particular Risks to Men

- Are there specific risks to men, including risk to sperm from study participation or effects of the study drug on sperm?
- Are examples of effective methods of birth control discussed? How long must such methods be employed?

Liability

- Are all reasonably foreseeable risks disclosed in the informed consent form, to fully prepare the participant and to minimize the possibility that an adverse outcome will occur that was not disclosed before the participant's enrollment?
- Does the informed consent form directly address liability in the event that the participant has an adverse reaction to study drugs or procedures? Is it made clear who will pay physician or hospital costs if an adverse outcome occurs, if compensation will be provided for long-term disability resulting from a severe adverse outcome, and how the study sponsor will decide whether or not a severe adverse outcome occurring during participation is related to the study? Is there clear statement of the participant's legal recourse if a study sponsor decides that a severe adverse outcome was not related to the study?
- Will the study sponsor promptly pay for all reasonable and necessary medical expenses the participant may have, including hospitalization?
- What are the provisions under which the study sponsor will pay for all reasonable and necessary medical expenses?

- What are the provisions under which the study sponsor will not pay for all reasonable and necessary medical expenses?
- Who will determine what is a "reasonable and necessary" expense?
- What criteria will be used to determine what is a "reasonable and necessary" expense?
- Who will determine whether or not an adverse outcome is related to study participation?
- What criteria will be used to determine whether or not an adverse outcome is related to study participation?

Alternatives to Participation in the Study

- Does the informed consent form describe the other medications, combinations of medications, and therapies that are available in standard care for the medical condition pertaining to the study?
- Does the form advise the study participant to discuss any alternative treatment options with his or her regular physician before enrolling?

Participant's Health Information

Health information typically includes the following: medical records, history and physical examination records, consultation reports, laboratory tests, x-ray and other diagnostic (imaging) reports, operative reports, discharge summaries, progress notes, questionnaires, interview results, focus group surveys, psychological surveys, psychological performance tests and other tests, photographs, videotapes, digital and other images, and tissue and/or blood specimens.

- Does the informed consent form discuss what parties may have access to any or all of the above information if authorized by the participant?
- How will the principal investigator determine whether the participant understands all elements of health information as they pertain to participation in the study?
- How will the principal investigator determine that the participant understands what he or she is authorizing by signing any statement regarding the release of information held by the principal investigator and research team?
- Does the informed consent form describe the risks should any element of health information fall into the wrong hands, the precautions that the principal investigator and research team will take to ensure that all elements of health information are protected at all times, and the fact that,

despite the best attempts at protection, health information may fall into the wrong hands?

Participation and Withdrawal

- Does the informed consent form clearly state that enrollment in this study is voluntary and that the individual may refuse to participate or may withdraw from this study (within parameters of safe withdrawal from the study) at any time without affecting the individual's relationship with or treatment by the physicians of the medical institution or other organizations in question?
- Does the form clearly state that the individual considering enrollment may consult a physician of the individual's choice who is not associated with the study to solicit that physician's opinion regarding participation in the study? In any case in which a participant's physician is paid a finder's fee for referring his or her patient to a study, this must be disclosed to the individual considering enrollment. (Any study in which a physician is both a recruiter for the study and would be advising his or her patient on enrollment is potentially problematic, because the patient could not be certain whether the physician was making a recommendation based on the individual's best medical care or the best interest of research.)
- Does the form state any and all requests that will be made of the participant if he or she decides to withdraw from the study before its completion? For example, will the individual be asked to complete an exit interview and/or physical examination or to evaluate the study in any respect?
- Does the form list the circumstances in which a participant may be asked to discontinue participating in or may be removed from the study?
- How will the individual be treated safely when discontinuing or being removed from the study?
- Will the participant, during the course of the study, be informed as new information becomes available which may or may not influence the participant's willingness to continue in the study? If so, how will this informing take place, and how quickly will the participant be notified of new information?

Costs to the Participant

- Are any and all costs to the participant and his or her insurance company specified?

Participant's Questions

- Does the informed consent form address how participants' questions will be answered before, during, and after participation in the study?

- Does the form provide the telephone number of the principal investigator for study-related questions? Who is the participant to telephone or otherwise contact if the principal investigator is out of town or not reachable by telephone?
- Does the form provide the telephone number of the IRB chair or other contact person for any ethical questions about the study, including complaints about the principal investigator's or research team's conduct of the study?

Participant's Rights

- Are the participant's rights listed and clearly explained?
- What materials are offered to the participant to further explain those rights?
- Is it prominently stated who can answer the participant's questions about those rights, and are those people's telephone numbers provided?
- Is there a clear statement that the individual does not have to enroll in the study and that his or her refusal to enroll will involve no penalty, loss, or loss of benefit?

Information and Its Role in Decision Making

Content refers to the definitions, terms, concepts, and descriptions of procedures in the informed consent form, and questionnaires connected with the study. *Presentation* refers to how the information is presented. The aim of both should be achieving the greatest clarity and readability possible.

All content should be translated into language the participant can understand. The IRB should not only criticize the text if necessary but should also suggest better wording of terms and concepts. As with the scientific protocol and the informed consent form, questionnaires that use technical language in regard to medicine, science, or other issues should be sent back to the principal investigator for translation. Usually, inclusion of technical language occurs because of ignorance or oversight, but IRB members must be alert for any attempt to hide issues or make them inaccessible.

The IRB should also analyze the way text appears on the page, as part of determining how to make the information most readable. The following elements of presentation can influence readability.[1]

Type Size. Is the type too small to allow for easy readability? Will the type size necessitate the individual's using reading glasses, and, if so, what will happen if the individual does not have reading glasses with him or her?

Type Font. Simpler fonts (styles of type) may be easier to read than an ornate font. Imagine reading the daily newspaper if it were printed in an ornate font.

Choice of Ink. Is the ink being used dark enough to provide good legibility? Are the colors of ink and paper—black or dark gray on white—appropriate for the serious information being conveyed?

Order of Text (Organization of Information). Individuals differ on the priority they give to various kinds of information. For example, some people will be more interested in information about the severe adverse outcomes at low probabilities, and others will care more about common side effects. Still others will weigh both equally. The more severe adverse outcomes and their chances of occurrence should be listed first, even if their probability is low. Low-probability events, when they do occur, tend to cause very serious outcomes, like stroke and death.

Line Spacing. People differ on how much spacing they like to see between lines of print. If too tight spacing causes overlapping of letters, there is a chance that an individual will avoid reading that material. Yet, too much spacing may make an already lengthy form too long for deliberation, even when read on several occasions.

Margins. Standard margins for an 8 ½-by-11-inch page are one inch on each of all four edges. If margins are too narrow or too wide, some individuals may experience difficulty when reading the document.

Headings. The IRB should pay attention to the headings and subheadings of all key areas. Is there a heading, on a separate line, to notify or alert the reader of a section's content? The lack of or inappropriate use of headings may be an attempt to hide material or to distract readers from focusing on particular sections (for example, the risk section).

Uppercase and Lowercase Type in Headings or Text. If the letters in the headings or in the text are uppercase only or upper- and lowercase, this may cause difficulty if the reader cannot quickly and correctly discern what is being said. Drawing attention to one area of text may mean drawing attention away from another area.

Line Length. The IRB must decide whether the length is appropriate for any older and more vulnerable individuals who will be asked to enroll in the study. Too many letters per line may make difficult reading for some individuals.

Bullet Points. Bulleting is useful in drawing attention to a particular text area or text sequence, but again, drawing attention to one area may draw attention away from another area.

Boldface Type, Box, or Summary to Highlight Important Points. Using boldface text, placing a box around text (particularly summary text), and providing summaries to highlight important points can help readers gain a better understanding of complex material. However, because focusing attention on one area may draw attention away from another, the IRB must ensure that appropriate emphasis is placed on each point.

No Watermark. The paper should not contain a watermark for two reasons. First, a watermark can decrease the readability of the section of text that is printed over the watermark. Second, if it is an institutional watermark, it may carry the same implications that the words "IRB Approved" may in the minds of some individuals; a potential study recruit may consider an institutional watermark as an endorsement of the study or of participation in it.

Illustrations. Relevant illustrations may prove very useful. Take the case of a participant randomly assigned to undergo one of the two different heart transplant procedures that are being studied. Anatomical illustrations can help the participant better understand the basic anatomy of the heart and its surrounding structures. They may also help the individual considering enrollment in the study to compare the two procedures with alternative interventions he or she could choose. The procedures section of the informed consent form for this study should contain both an illustration of the anatomy of the heart under each procedure and a verbal description of how the two procedures differ. Further illustrations can show the different physiological effects each procedure will have on the participant's body over time. Selective, well-designed, well-drawn, and well-labeled illustrations, both anatomical and physiological, when used judiciously, can be an important component of a well-designed informed consent form.

Financial and Legal Liability for Adverse Outcomes

In the IRB's initial review of an informed consent form, it makes an assessment of issues related to compensation for damages that participants may incur while participating in research, the legal liabilities involved in participation, and whether or not the study sponsor or the medical institution will pay the costs of adverse outcomes that occur to participants. *Compensation* refers to the compensation of participants for the personal costs of participating in a study or of harm done to them. *Legal liability* refers to the responsibility under the law of the various parties involved in the study in regard to injury or adverse outcomes incurred during participation in a study.

The IRB member must make certain that there are clear statements concerning legal liability, particularly if the participant may sustain an adverse outcome from participation in the study. Given that there may be disputes as to what caused the adverse outcome if it occurred during study participation or even afterward, the IRB must ensure that participants understand their right to bring suit to recover damages from a study sponsor or related party. The informed consent form needs to make clear:

- who will pay for hospitalization for a study-related condition, particularly hospitalization involving an adverse outcome of the study;
- who will decide if the adverse outcome is study related;
- what recourses are available to the participant should the study sponsor decide that the adverse outcome is not study related;
- who will pay for the drug after the study is finished if the study drug was found to be beneficial to the participant; and
- who will pay for long-term costs of study-related permanent injury.

The IRB member must look for exculpatory language that might be interpreted as the participant's giving up rights related to any injury due to negligence by the research team or staff or that might be interpreted as protecting the principal investigator and/or study sponsor from responsibility to the participant. The IRB must ensure that all exculpatory language is removed.

If the sponsor and principal investigator have not done so, the IRB must provide forethought regarding the costs to the participant for his or her involvement in the study. Some sponsors and principal investigators may try to avoid paying any of the participant's costs and to leave the burden of costs of participation and of adverse outcomes on the shoulders of the participants.

The IRB must protect the participant by ensuring adequate disclosure of the risks, clear accounting of costs for injury, and communication of the fact that the participant has the right to sue for damages from injury incurred because of participation in the study. If these things are not communicated clearly, the potential participant cannot validly consent to participate.

Changes in the Consent Environment

Informed consent is a continuing process. The form is signed at a particular time but the decision it represents can evolve over time. As new adverse outcomes are reported and data trends are observed, the balance of benefit and risk weighed in the original decision may shift. The study itself may be modified, suspended, or terminated (in which case the participants should be informed of why the study was stopped).

The informed consent form may need to be modified and reviewed by the IRB. Participants in the study will need to read and sign the new informed consent form.

The Informed Consent Session

More and more IRBs are asking principal investigators and research teams to make certain that they obtain from individuals the best possible informed consent to participate in the research, and this goal is furthered by face-to-face discussions with potential participants before they sign the form. The informed consent session allows an individual considering study participation to discuss aspects of the research trial and study participation one-on-one with the principal investigator or his or her designee. In the vast majority of research studies the session will be one-on-one. The session is a great opportunity for explanation, but material must not be omitted from the consent form under the assumption that it will be covered in the consent session.

The ways in which principal investigators and research teams conduct informed consent sessions and the responses to questions are as varied as the individuals engaged in the sessions. IRB members may observe an informed consent session at any time after a study has been approved by the IRB and is operational. Many useful suggestions can result from the review of informed consent sessions. Interviewing people is a surprisingly complicated process, and anyone can benefit from training in this skill. Principal investigators and interviewers can be observed for strength or weakness in the recruiting process and in the way they answer questions individuals have after reading the informed consent form. They can be instructed in how to improve their communication skills to more optimally convey risk information to potential participants. Because research teams change over time and new members enter, the process of reviewing informed consent sessions needs to be ongoing.

7 Recruitment, Selection, and Compensation of Study Participants

Institutional review boards should survey all strategies used around the country to recruit study participants, with the goal of asking principal investigators specific questions about how individuals are being recruited.

Box 7.1. Questions the IRB Should Ask about Recruiting of Participants

- Who will be asked to participate?
- Who will contact potential participants?
- How will potential participants be contacted?
- Will off-site recruiting be done?
- How will the IRB counsel the principal investigator on contacting participants without using uniquely identified patient data?
- Does the institution in which the IRB resides provide an area, newsletter, or website where research opportunities can be announced?
- Which means of contacting potential participants does the IRB wish to allow and which to prohibit?

What Type of Recruitment Will the IRB Allow?

The IRB will face many challenges in determining what to allow in study recruitment. It will need to review and approve or reject all recruitment strategies and materials before their use. The principal investigator must specify both how and where recruitment and advertisement will take place.

Recruitment by Brochures, Websites, Newspaper Advertisements, and Posted Advertisements

Recruitment information can be made passively available in clinics and it may be passed out to all patients attending a particular clinic. In addition, medical institutions can have a website advertising the research opportunities that are available at the institution, and this web address can be posted in clinics. The medical center can advertise research opportunities in its newspaper. Advertisements can be posted around the hospital in places frequented by potential study participants, giving telephone numbers they can call to find out more about the study.

A poster or flier about a study should state basic eligibility criteria (e.g., age, having been diagnosed with the condition being studied), any benefits that might be received, basic risks, and contact information.

Brochures and longer advertisements should state explicitly whether the study compares two drugs, compares a drug and placebo, or has another study design. A brochure or advertisement should not be intentionally vague to attract potential participants only to have the truth come out later. Principal investigators must practice honesty in advertising at all times, and IRBs must see that these materials provide enough information.

The question of how much should be said about risks in an advertisement or brochure is intriguing. A complete list of risks would never fit. I recommend that, at least, there be a statement to the effect that "the principal investigator or his or her designee will fully discuss all risks of participation in the study."

Recruitment by "Cold" Telephone Calls

"Cold" calls are telephone calls that are unsolicited and unexpected. A cold call indicates that someone has gained access to information about the person being called and his or her medical condition and believes the individual is a candidate for a research study. Thus, cold calls may imply a release of medical information to which the individual did not agree or that he or she may at some point have signed a release form without reading or fully understanding it.

An acceptable alternative to a completely cold call is to make an initial contact with patients when they are visiting a clinic and ask if they would be willing to be called to discuss the research study and whether they would like to participate in it.

Recruitment by Letter

Recipients of a recruitment letter mailed to the individual's home may well ask: How did they obtain my name and learn my medical condition? How could they gain access to this information without my signed agreement? Why did the medical center or my doctor release my name and disclose my medical condition? It is quite correct to pose these questions to one's doctor or the institution that sent the letter.

Often, the mailing includes a postcard for the individual to return, indicating whether he or she is interested in further contact about the study. The IRB must decide what they will permit the postcard to say. The text should not disclose the substance of the research, to protect an individual from being linked publicly with the disease or medical condition being studied.

Most such letters instruct the recipient to mail the postcard back if the individual is interested. Some letters say that if the postcard is *not* returned it will be presumed that the individual *is* interested. The latter approach is problematic in terms of honesty and appreciation of the individual's time.

Problematic Techniques of Acquiring Patients' Names for Recruitment

Recruitment techniques should not involve any intrusion into patients' medical records without their specific permission. Any attempt to gain access to medical records without the patient's explicit approval may violate the confidentiality of patient data if the data are uniquely identified. Imagine a case in which a patient with a particular disease is cold called by a principal investigator or research team member and asked to participate in a study related to that disease. What will the researcher say when the patient asks, "How did you get my name to ask me about this disease?"

Who Will Contact Potential Participants?

Who will be making further contacts with potential participants? Volunteers may be useful but will need to be trained in how to contact a patient and what to do with any questions the patient has regarding the study. Trained volunteers may make contact, but it takes well-trained personnel to handle patients' questions.

Selection of Study Participants

The guidelines for the selection of human participants in research that the IRB provides to its principal investigators should specify that there should not be an unfair burden placed on one group of patients (e.g., a small number of people with an unusual disease being recruited for many studies) and that there should not be an unfair exclusion of patients from participating in a study. Factors brought to bear on such decisions include the appropriate selection of participants across all socioeconomic categories, ages, genders, races, special populations, levels of education, and geographic and regional characteristics.

Not all individuals being approached to participate in a research study will have the same decision-making capacity. Unfortunately, there are no mutually agreed upon scientific tools used to validly measure decision-making ability for research participation. As I will discuss in Chapter 14, decision-making capacity is a point of concern in many studies, particularly those involving conditions or processes that affect an individual's ability to perceive, cognitively process, understand, and/or communicate information.

Compensation of Study Participants

Compensating participants (e.g., for the costs they pay out of pocket while participating in a study) used to be frowned on as possibly inducing some individuals to enroll in research studies. However, today it is recognized that time spent in research has certain costs (e.g., time away from work, time in travel, gasoline and other travel-related costs, meals, even overnight accommodation for certain protocols). Therefore, the IRB must address the issues of whether participants should be compensated and, if so, how much compensation is reasonable. When considering these issues, the IRB will need to know if the participant might be asked to return after the initial study involvement, to complete quality-of-life surveys, have additional medical histories taken, exams or tests done, or, in the case of genetic studies, to have genetic histories taken.

Each time point and time interval of participant involvement may be associated with a time or financial cost for the participant. The IRB member must ask whether the participant will be compensated for them or if the principal investigator simply assumes that the participants will absorb the costs. The IRB member must ask when it is inappropriate for the participant to pay out of pocket. All of the possible expenses and who will pay for them must be clearly stated in the informed consent form.

In addition, the IRB will have to decide how much compensation is "too much." Too much compensation may be interpreted as inducement to participate in a research study. The informed consent form must state the amount of payment and schedule of compensation, if any, for participation in the research study.

Study sponsors and principal investigators may try to overpay participants, to procure their willingness to stay in a study over time. The IRB members need to weigh these competing interests involved with participation in research.

The IRB should consider whether the schedule of payment, not just the amount paid, may be an inducement to participate in the study. Consider, for example, the comparative ramifications of a lump sum paid on entering the study versus a pro-rated payment schedule versus a lump sum paid on completing the study. Although the lump sum paid at commencement may be considered an unfair inducement, a lump sum paid on completion may mean that the participant will be fully reimbursed for gasoline, food, or even room and board. A lump sum payment at the end of the study would certainly be an inducement for a participant to remain in a study longer than if he or she received payment after each study appointment or procedure. However, a lump sum paid at the end of the study might also be an unfair inducement to continue, if a participant wanted to withdraw for a legitimate reason.

8 Research involving Questionnaires and Surveys

Research in both the behavioral and the medical sciences often includes questionnaires and surveys. For example, much medical research involving organ transplantation, severe cardiovascular disease, severe cancer, and severe pulmonary disease includes a quality-of-life questionnaire or depression questionnaire. The assessment of patients' opinions and preferences regarding quality of life is key in a large number of behavioral science and medical science studies.

Because of the high level of attention being paid to research on human participants in the United States today, the challenges in designing a questionnaire or survey are just as important as those in any other type of study. Some issues regarding such studies to which IRBs must attend are given in Box 8.1.

Box 8.1. IRB Concerns regarding Questionnaire- and Survey-Based Research

- Number of questionnaires over the course of the study
- Length of time required to complete each questionnaire
- Content of each questionnaire
- Reason each questionnaire is being used in the study
- Issues related to stress, negative emotions, and negative mental states generated by questionnaire or survey studies

The IRB needs to analyze the questionnaire for both its information content and how the information is presented. The IRB member must read each question, assess its phrasing, make a judgment about the sensitivity of the subject matter, and make a determination of the overall quality and suitability of the questionnaire or survey. If the IRB has questions about study participants' sensitivity regarding topics being raised and questions being asked, the IRB may need to consult local, regional, or national experts to determine the safety of the questionnaires for use with particular groups of people.

In most questionnaire or survey studies, the potential participant should be informed of:

1. the number of questionnaires or surveys that the participant will be asked to complete;
2. the timing of the questionnaires and whether they are to be completed in one sitting, in more than one sitting on the same day, or on multiple days;
3. if multiple days will be involved, the detailed schedule of the days, number of sessions, number of questionnaires, and the time required to fill out each questionnaire.

IRB members must decide whether the principal investigator has accurately estimated how long it will take an average participant to complete the questionnaire(s) and, if they find adjustment called for, should instruct the principal investigator to make appropriate changes to the protocol and the informed consent form.

Challenges in the Construction of Questionnaire Studies

The IRB must assess the hypothesis underlying the study and help the principal investigator develop the best questionnaire or survey that can be constructed to answer the questions of the study hypothesis. For example, if the questionnaire is asking participants for preferences, is it clear what is being asked? Can the IRB member sharpen the phrasing of the questions? Or take the case in which participants will be asked to weigh risk and benefit in a real or hypothetical treatment decision. If the principal investigator is phrasing the choice as a gamble, is that the best approach? Some individuals may not like having treatment decisions described in terms of gambling, because of the moral issues associated with gambling. Could the principal investigator achieve the same purpose by using the word *trade-off* instead of *gamble?*

In the behavioral sciences, the challenge is to determine what information about the survey can be disclosed to the participant without compromising the study results. Sears, Peplau, and Taylor state, "The researcher has an obligation to tell potential subjects as much as possible about the study before asking them to participate. Subjects should be informed about the research procedures, any risks and/or benefits of the research, and their right to refuse to participate or [to] withdraw [their participation] during the research without penalty."[1] Some IRB members would go a little further: the individual has the right to understand as much as possible about the study before making the decision whether or not to enroll, and the researcher has an obligation to support the individual's right.

Sears, Peplau, and Taylor note that although "the requirement of informed consent sounds quite reasonable, [the requirement] can sometimes create problems for social psychologists." For example, it "may be important not to tell subjects the true purpose of the study, to avoid biasing their responses. Even in the simplest research, subjects are rarely told the specific hypotheses that are being tested."[2] Some IRB members may be sympathetic to this position, and others may disagree.

Challenges of Informed Consent

The researcher's notion of informed consent is not necessarily that of the IRB. The *Code of Federal Regulations* states that, with regard to informed consent in research on humans, a basic element is that "in seeking informed consent the following information shall be provided to each subject": "A statement that the study involves research, an explanation of the purposes of the research, and the expected duration of the subject's participation, a description of the procedures to be followed, and identification of any procedures that are experimental."[3]

In the informed consent form for a questionnaire study, the principal investigator must answer all of the questions in Box 8.2.

Box 8.2. The Informed Consent Form in a Questionnaire Study

- Is the questionnaire or survey anonymous?
- If the questionnaire is not anonymous, has the principal investigator disclosed to the participant that it will be identifiable with him or her, and does the informed consent form specify who will be able to gain access to the questionnaire data, for what purposes the data will be used, and how long the data will be kept, among other issues?
- Could any issues come up in the questionnaire that will have to be communicated to the participant's primary care physician? For example, a participant may be asked to complete a depression screen, and positive results of depression will need to be communicated to the participant's health care provider.
- Will the questionnaire be made part of the patient's medical record? If so, whose responsibility will it be to record laboratory and study results in the patient's medical record?
- Where will the questionnaire data be kept, and will it be under lock and key?
- Will any nonanonymous data (that is, data with unique identifiers, such as name or hospital number) be stored in a computer? If so, how will that data be protected?
- Is it possible that, no matter how well kept the data are, someone outside of the research team could gain access to the data and use them for deleterious purposes? If so, is that explicitly stated?

- Does the questionnaire ask for sensitive or private information, for example, about illicit behaviors? If so, how will the IRB handle this sensitive information if others want to access it? Will the IRB insist on reporting abuse to the appropriate parties?

The IRB member must make certain that new medical information discovered in the course of a questionnaire study is relayed immediately to the participant's physician so that appropriate action can be taken to minimize damage to the participant, but the participant must be informed that this will be done and when it is being done. If abnormalities are found during a study, these abnormalities must be discussed with the participant and managed without delay, either by the research team or by conveying the information to the participant's physician.

Recognizing Potentially Problematic Questionnaire Studies

A questionnaire study is no less risky than other types of research (e.g., those involving invasive procedures). IRBs often underestimate the risks posed to participants by studies involving questionnaires and surveys. Principal investigators need to be prepared for adverse reactions to questionnaires just as for adverse responses to medical interventions. The IRB must counsel the principal investigator to be ready with explicit responses by research staff and with persons trained to help the individual experiencing the reaction. Provision for help should be well thought out before the study is undertaken.

Questionnaires and surveys may elicit bad memories and even irrational actions in someone asked to complete the survey. For example, studies inquiring about the death of a loved one or considerations of a genetic disease may be quite disturbing. Questionnaires and surveys can even precipitate post-traumatic stress disorder. Other potential problems include the handling of sensitive data that participants may not want to have released, like incriminating data about crimes and data concerning the genetics of a family that may have an impact on not only the participant but also the participant's blood relatives past, present, and future, who may not even be aware of the participant's involvement in a genetic study.

Other questionnaires may contain questions that individuals consider private and confidential and will refuse to answer or become irate at simply being asked. Very sensitive issues arise in the case of questionnaires involving individuals who have been or continue to be the objects of abuse. In the case of past abuse, how will the participant be counseled when the memories of abuse are called up during completion of the questionnaire? In the case of ongoing abusive relationships, how is the participant to be helped? Clearly, the principal investigator must develop a framework for helping such participants and must deliberate about the best ways to involve counselors to help these participants cope with the issues raised by memories of events. The IRB must be aware of whether the state where the study is being conducted mandates the reporting of abuse when it is discovered by health professionals. If so, then the IRB must assure that such reporting is done by the principal investigator or a designee and that the participant knows that this will occur.

Declaring that "if abuse is discovered, it will be reported" might deter some people from participating and might inspire others to participate. But truth in disclosure is essential if individuals are to trust in research on humans and to trust the study sponsors and investigators who conduct it. As always, it is the individual who voluntarily agrees to bear the risks of that research participation and who needs to understand the full range of risks he or she is asked by society to bear.

The emotional and psychological consequences to the study participant should be considered in all potentially sensitive areas of research (see Box 8.3). The IRB and principal investigators should examine the wide range of potential negative consequences in sensitive areas. There should be not only on-site counselors or health care providers present to help participants but also access to specifically trained counselors and providers after the questionnaire session. Participants need to receive specific information, including telephone numbers of counselors and names of providers to contact when the participant has problems that need attention that night, the next day, or at a future time.

Box 8.3. Potentially Sensitive Areas in Questionnaire Studies

- child abuse
- elder abuse
- spousal or partner abuse
- drug abuse
- post-traumatic stress disorder

Finally, the setting in which the survey is taken can be problematic. For example, in the case of a workplace survey, unions may require notification of the employer's intention to survey workers and specification that the employee will be allowed time off work to complete the survey and that employees who decline to complete the survey will not suffer repercussions for their refusal.

9 Protection of Participants' Privacy in Research Data and Specimens

So far, this text has focused on one key aspect of the protection of study participants: their well-being during and after participation in the study. The IRB must also make certain that all *data* derived from the participant are protected and held confidentially, to protect the participant's privacy, and that all *specimens* the participant has donated for use in present and future research projects are managed optimally for the participant's privacy. With genetic samples, the privacy protection must extend to the participant's past, present, and future generations.

The IRB must keep in mind the need to protect research data at all times and the fact that data may be inadvertently released to a third party. IRB members need to consider all of the consequences of inadvertent release of information, which may harm a study participant and/or his or her relatives and future generations.

Basic questions regarding the protection of research data in general are shown in Box 9.1. The IRB must understand the types and layers of protection that principal investigators and research teams must place on data and that no protection, however careful and well intentioned, is complete and infallible.

Box 9.1. The Protection of Research Data

- How will the data be protected?
- Who will have access to the data?
- In what form will individuals have access to the data?
- What are the chances that the data will be inappropriately accessed, and how will any and all inappropriate channels of access be blocked?
- Will the data be identified uniquely or by a random study number?
- If the data will be identified by a random study number, will there be a code book linking the study number with a participant's unique identifier (e.g., hospital record number, home address)?
- If there will be a code book, how will it be protected and who will have access to it?
- Do the study and its results involve the transfer of data from one institution to

another? If so, how is the participant's anonymity to be protected in the way the data will be coded for transfer?

—To where might the data be transferred?

—Who is responsible for the safe storage of transferred data at all sites and at each site?

—What are the conditions of safe storage of transferred data at the other sites?

—How is the safe storage of data at the other sites to be checked?

The planned means of storage and any transmission of data must be examined by the IRB for each study. Basic questions regarding data storage are shown in Box 9.2.

Box 9.2. The Storage of Research Data

- What data will be stored?
- Where will the data be stored?
- Will the data be stored
  —physically (e.g., as lab sheets in a filing cabinet)? If in a locked place, who will have the key?
  —electronically (in a computer)? If so, what security will be used to protect the data?
- Who will have access to the data? This question is answered by obtaining the following additional information:
  —names of persons with access to uniquely identified data
  —positions of the individuals in the principal investigator's research team or study sponsor's organization
  —reasons each individual needs to have access to uniquely identified data for research purposes
  —reasons each individual needs to have access to uniquely identified data for nonresearch purposes
  —level of risk if any uniquely identified data are misused in any context
- How will the data be protected?
- Who on the research team is in charge of protecting data?
- Who will be protecting the data when the original research team no longer exists?

Uniquely identifiable data are even more sensitive. All efforts must be made to ensure that any papers bearing uniquely identified information are shredded and placed in secure containers when disposed of. The IRB needs to emphasize optimal vigilance over uniquely identified data that can be subject to theft or inadvertent removal once it is out of the direct control of the principal investigator and research team.

It is difficult enough for the IRB to ensure that all data are being protected within its own institution; however, once data leave the institution, it is virtually impossible to ensure that the data will be protected in the manner the IRB believes to be safest for protecting participants. In the case of uniquely identifiable data (if in fact the IRB has allowed such data to be collected), the IRB may mandate that it never leave the institution, because of the threat of theft of CDs, diskettes, laptops, thumb drives, and the like. Questions about transmission of data are shown in Box 9.3.

Box 9.3. The Transmission of Research Data

- At time of transmission, will the data still have some remnants of unique identification? If so, why?
- Will data be transmitted only between members of the research team?
- Should transmission of the data outside of the institution be allowed?
- If the data will be transmitted outside of the institution, does the participant give explicit permission by signing the informed consent form?
- How will the data be transmitted within or outside of the institution? What procedures are in place to protect the data during transmission?
- What does the informed consent form tell the participant regarding how the data will be kept and the limits of the protection of data?
- Who gave permission for transmission outside of the institution?
- Who will have access to the data outside of the institution?
- What are the limits of protection?

Just as data must be carefully protected, specimens of blood, cells, tissue, and organs collected as part of a study must be handled in such ways that the participant's privacy is protected. The IRB must ask the same sorts of questions it asks regarding data protection. However, because all of this material carries genetic information, the challenge and responsibility are greater. Genetic studies can involve data about genetic mutations within families, and therefore the significance of genetic study findings extends beyond the individual who agrees to donate bodily specimens.

The use of specimens for genetic studies poses special problems to the IRB in its role as the protector of study participants, because the potential risks extend beyond the individual signing the informed consent form and/or agreeing to donate specimens and may involve past, present, and future generations. The IRB needs to know of any potential risks, especially loss of employability or insurability, for the participant and the participant's current and future blood relatives. Because threats to employability and insurability owing to genetic diseases are not currently adequately addressed by law, the IRB's scrutiny of this issue must be intensive and its decision making must focus on the protection of all who may be affected, not only the individual considering donation and study participation. How this matter is presented in the informed consent form is extremely important. In genetic research, the study participant's blood relatives in past, present, and future generations become nonconsenting participants.

When evaluating the handling of biological specimens, the IRB must consider at least the questions in Box 9.4.

Box 9.4. The Protection of Genetic Material

- Is the participant agreeing to the use of tissue in the laboratory of a single investigator?
- Is the participant agreeing to its use by multiple investigators in a single medical institution?
- Is the participant agreeing to its use also by investigators in neighboring and interrelated institutions?
- Is the participant agreeing to its use by universities and the Department of Veterans Affairs medical centers, but not by private industry?
- Will any of the tissue to be sent to private industry (for example, product manufacturers)? If so, who will authorize the transfer?
- Will the sample remain in the country of origin, or will it be stored elsewhere? Why would any sample leave the original medical research institution?
- When does a sample derived from a human study participant contain genetic information?
- Will DNA contained in a sample be extracted as part of the study, or will tissue be stored for future DNA extraction?

- Is the specimen to be used and/or stored for future studies that may be of a genetic nature?
- Under what conditions does the release of genetic information pose a risk to the participant and/or his or her genetically linked relatives?

The scope of protection of future generations goes beyond the regulatory framework for current research. The IRB must make decisions regarding research and the protection of human participants related to study hypotheses that have not even been conceived.

Clearly, stored tissue may outlast the lifespan of the principal investigator and research team. Who has the obligation of protection when everyone on the research team has died? Remember that, under conditions in which the participant is uniquely identified, tissue = genetic information = risk.

The problem in protecting study participants is greatest when stored genetic material includes a unique identifier that links it to an individual. Except in the case of rare diseases, if stored genetic material is not uniquely identified, there is less of a problem.

Identification of Samples

The obvious unique identifiers in a medical record are the patient's name (unique in some but not all cases), hospital record number, and Social Security number. However, there are other ways a patient's identity can be discovered. For example, if a hospital has only one operating room in use at a specific time and the surgical record contains the patient's diagnosis, the type of surgery performed, and the date and time of the surgery, someone who has access to the hospital database could identify the patient. Or if a patient is the only one with a rare medical condition or disease in a medical center, someone with the sophistication and access to search the hospital database for that specific disease can gain access to further identifiers, such as name and Social Security number.

In a database, unique identification can be discouraged by separating data into subgroups, for instance, by recording not specific ages but age ranges (e.g., all participants between the years of 90 and 100). However, in some small clinics or hospitals, there may be only one participant between the ages of 90 and 100. For the person who is willing to probe, the simple identification of an age range may sometimes be enough to allow the identification of a participant. Additional solutions are needed to prevent parties to whom the participant has not given explicit permission from obtaining the participant's information because of *potential* identifiers. These additional solutions should prevent the sharing of unique data for which the individual has not given explicit permission.

Explicit permission entails an informed consent form and process that educates each participant about the uses and misuses of uniquely identified data, especially the risk of loss of employability and/or insurability based on the fact that an individual has a specific medical condition or disease.

Future Research

There must be explicit discussion, in a section of the informed consent form labeled "Notification about Participation in Future Research," regarding the principal investigator's intentions with respect to future research. The participant needs to understand what participation in future research might involve and what limitations the participant might want to place on use of his or her information in future research. Some IRB members argue that any notification about future research is fundamentally flawed in that it is difficult to delineate the limits to be placed on it.

All sharing of uniquely identified information with parties not involved in the original study is problematic unless the participant has granted specific and explicit permission for such sharing. Once any data that uniquely identify the participant leave the site of the original study, it will be impossible for the IRB to oversee how the data are being used and what further sharing of the data is taking place. Each IRB must directly address and restrict the sharing of data unless the participants in the dataset under consideration have discussed this and granted explicit permission.

It is important to note that there are ongoing debates about certain types of data that can never be collected again and that therefore may need to be shared with parties not involved in the original study. For instance, if a patient with a specific genetic disease dies, that individual cannot voluntarily donate samples for research purposes. If there is no advance directive donating the body for specific research purposes related to genetic research and the family does not consent to a donation, there will be problems obtaining further genetic material for continued research unless a relative agrees to be a study volunteer. Another reason to share information may occur if a principal investigator realizes that he or she does not have the relevant expertise to analyze some of the data collected. Here, the principal investigator may ask the IRB for permission to bring in new parties to help analyze the data. The IRB may need to examine who the parties are and what their interests are and to follow this development over time.

Even when data or samples are not uniquely identified, problems can arise if principal investigators share the dataset of patients with a specific disease with principal investigators at other institutions without obtaining specific permission from each patient. What would a patient think if he or she were contacted by another medical institution about participating in a study involving the specific disease or condition the patient has? The first question would be, Who allowed someone at a different institution access to my private information without my permission? The IRB must be aware of strategies that some principal investigators and study sponsors will use to circumvent obtaining explicit permission to share information. These strategies include using general permission statements, for example, "permission to contact the patient in question about future research studies" when nothing is specified regarding these future research studies. Imagine an informed consent form that contains the above statement about future research studies positioned in a nonexplicit fashion within the body of the form. If the patient signs that form, should the principal investigator be allowed to assume anything regarding what a participant has agreed to?

Will Data or Specimens Be Sold?

At present, without clear regulation, there is a high potential for misuse of specimens in genetic research. In some genetic studies—involving donation of blood and tissues—investigators have asked participants' permission to sell their specimens, without any explanation of who the purchaser might be or what use might be made of the specimens. The participant might be told that the donated specimen will be used for "*research purposes only,* not for sale." Each informed consent form must explicitly state what uses might be made of the donated uniquely identified tissue.

The IRB must be vigilant to ensure that no sale takes place and that specimens remain protected for selective use in research projects with the specific and explicit types of direction that a participant gives to a principal investigator. This direction includes the research uses that can be undertaken with the donated tissue, who can conduct the research, and for what purposes. However, the IRB must recognize that at present it is not always clear whether genetically linked relatives agree with the participant's decision making regarding the specimens. Unless there is a procedure for informed consent by the patient's entire family for use of their shared genetic material in the research enterprise, there is a risk of unauthorized use. The term *research enterprise* is used here to include the financial issues that are inextricably liked to the use of human tissue; products can be developed based on the research derived from these donated tissue specimens. The specimens may not only be used in present scientific studies but may also be stored in tissue storage banks for future research studies that have not as yet been formulated.

Tissue Storage Banks

Researchers and their institutions face many challenges in developing and maintaining tissue storage banks for future research. An institution must be certain that it considers the following questions:

- What *limits* should be placed on the types of research for which scientists could be permitted to use the specimens?
- What genetic scientific hypotheses should be considered *reasonable* scientific questions to be addressed by research on humans?
- Who will decide and what strategies will be used to determine acceptable versus unacceptable research and hypotheses?
- How should statements disclosing future uses of stored tissue be worded?

It should be remembered that the issue of consent to use tissue is completely independent of the question of whether the specimens are uniquely or non-uniquely identified. Which genetic scientific hypotheses an individual or genetically linked family may consider reasonable to participate in is independent of whether or not the data is traceable to the individual or family.

The Consequences of Released Information

The permission given by a study participant and by family members to have their tissues, stored or otherwise, used in genetically based studies may have an impact on all of their contemporary and future blood relatives if the samples are allowed to be uniquely identified. These impacts might include threat to health and life insurability and employment discrimination based on the released information. Although protecting a participant's information is foremost in the minds of most principal investigators, the fact is that this protection, however conscientious and thorough, cannot be guaranteed, because of potential access, deliberate or accidental. Informed consent forms must be structured in such a way that participants clearly understand the consequences of inadvertent release of data.

HIPAA Privacy Regulations in Research

The federal Health Insurance Portability and Accountability Act of 1996 (HIPAA), which was implemented in 2003, emphasizes the protection of patients' privacy in regard to their personal health information. The IRB must understand HIPAA regulations, because they apply to all research data having to do with human subjects.[1]

Uniquely Identified Data

HIPAA emphasizes the need to obtain a participant's authorization for the use of his or her health information in research studies if in the research the individual will be identified by any one of eighteen unique identifiers (see Box 9.5).[2] The IRB must bear in mind these identifiers as it assesses the ways in which a participant can be identified with research data.

Taking account of these possible ways to identify individuals is a good beginning to the protection of privacy, but the range of information used in research extends beyond a participant's medical record. This being the case, a key component of IRB review of each research study must be systematic examination of the scientific protocols and informed consent forms documenting how the principal investigators and study sponsors will protect the participants' privacy and the privacy of data in the study, and also a continual review process initiated and maintained by the IRB during the study.

Privacy concerns would extend to a principal investigator's attempting to obtain contact information from not only participants but also participants' relatives. The IRB should carefully consider the necessity of such contacts and the risks to the privacy of the participant's relatives. The participant is the consenting study volunteer, not the relatives.

Let us consider an example concerning the genetics of heart disease. When a principal investigator wants to study heart disease in families, the proper way to do so is to obtain consent from each family member on an IRB-approved informed consent form accompanied by an informed consent session. The fact that one individual gives consent to participate says nothing about other family members' willingness to participate. Indeed, certain family members may object to research participation by any family member, and such views must be discussed within the family. An individual cannot give permission for another to participate in research unless he has been assigned legal decision-making authority for that person. The principal investigator must seek the consent of each family member with full discussion of risk and benefit to not only the individual but also the family.

IRBs should be concerned with protection not only of study data but also the participants' personal data (including contact data). Every time a principal investigator or study sponsor asks for a participant's personal data, the participant has a right to know all the uses the principal investigator or study sponsor may make of the data. The request for contact information may include a re-

Box 9.5. HIPAA's List of Unique Identifiers

- Names
- All geographic subdivisions smaller than a state, including street address, city, county, precinct, ZIP code, and their equivalent geocodes, except for the initial three digits of a ZIP code if, according to the publicly available data from the U.S. Bureau of the Census:
  —the geographic unit formed by combining all ZIP codes with the same three initial digits contains more than 20,000 people; and
  —the initial three digits of a ZIP code for all such geographic units containing 20,000 or fewer people is changed to 000.
- All elements (except year) of dates directly related to an individual, including birth date, admission (to hospital) date, discharge data, date of death; and all ages over 89 and all elements of dates (including year) indicative of such age, except that such ages and elements may be aggregated into a single category of age 90 or older
- Telephone numbers
- Fax numbers
- Electronic mail addresses
- Social Security numbers
- Medical record numbers
- Health plan beneficiary numbers
- Account numbers
- Certificate/license numbers
- Vehicle identifiers and serial numbers, including license plate numbers
- Medical device identifiers and serial numbers
- Web universal resource locators (URLs)
- Internet protocol (IP) address numbers
- Biometric identifiers, including fingerprints and voiceprints
- Full-face photographic images and any comparable images
- Any other uniquely identifying number, characteristic, or code

quest for the names and telephone numbers of participants' family members or significant others who have not been informed regarding their potential role in a study. Each participant must be offered opportunities to limit the uses of his or her personal information.

The IRB must ask:

- Who is to obtain permission for the giving and use of the names and contact numbers in a research project?
- What, precisely, are the names and contact numbers to be used for?
- Are the names and contact numbers really needed in the research study?

If the names and contact numbers are to be used for recruitment for future research studies or given to product manufacturers for purposes other than research, then the principal investigator and study sponsor must inform the IRB of all uses of the data to ensure that the IRB can verify that appropriate measures are in place to protect the privacy of participants and their relatives.

To assess a study's planned and ongoing adherence to HIPAA regulations, the IRB should have a working knowledge of the terms described below. These terms are important in understanding the arguments that study sponsors, product manufacturers, and principal investigators use to explain why they wish to obtain private information. The IRB's goal is to protect research participants, not to foster goals of study sponsors, product manufacturers, or principal investigators. The IRB will need to work out the strategies they will use in applying HIPAA rules to the protection of participants.

Deidentification of Data. For data to be "deidentified," according to HIPAA in its present form, the eighteen specific identifiers must be removed from records.[3] Deidentified health information does not identify an individual uniquely. The IRB must be assured that no reasonable basis exists to believe that the information could be used to identify an individual.[4]

Statistical Deidentification. If the institution in which the IRB sits chooses to use statistical deidentification as a means of deidentifying data, the IRB will often need help in ensuring that acceptable statistical techniques of deidentifying are being used. The IRB and the institution should consult an unbiased statistician to obtain properly qualified statistical expertise. After an approach for deidentification has been selected, the IRB must, at appropriate times, revisit that approach to see if new, more secure means might be preferable. The IRB needs both to understand the processes of protection and to determine whether and how risk is limited by them and to what extent risk of breach of privacy remains.

Limited Data Sets. In limited data sets, decisions are made on which data elements are to be included (or not) and the amount of specificity of the values of those data elements. If unique identifiers are eliminated from those data sets, there is the least chance of identification of individual participants. The goal is to minimize or eliminate the chance of tracing the data to a unique participant at any future time. However, if there is in existence a full data set from which data have been selected, this full data set is still in need of protection.

Computational Methods of Deidentification. In contrast to limited data sets, computational (computer-based) solutions are argued to render data "sufficiently anonymous" and can be used to make decisions on a participant-by-participant basis. Such computational approaches require that data be reviewed with the purpose of recognizing which participants are uniquely identified, then only their values are modified.[5] If all unique identifiers are not stripped from the data set before computational methods are applied, a full (or less full) data set still exists and will still need protection against loss, misuse, unauthorized access, disclosure, and alteration.

Part III

The Continuing Work of the IRB

10 The Ethical Issues of Informed Consent

To protect human participants in research, the IRB must perform many tasks. One of them is systematically to review over time the ethics of the study in regard to informed consent and the protection of study participants. In doing this, the IRB must distinguish clinical research conducted on human subjects from the clinical care of human patients. This is more difficult than it at first might seem, because participants and investigators attach different meanings to the terms *research* and *clinical care*. This chapter explores informed consent in clinical research and describes the differences between informed consent in research and informed consent in clinical care.

Research is the development of new knowledge for future populations. From this basic perspective, one might define research on human participants as the development of new knowledge for future populations based on the voluntary participation of humans. My own definition would include the notion that this development of new knowledge is conducted within an environment focused on the best protection of the human subjects. For our purposes, we will say that *clinical* research is the development of new knowledge for the care of patients and includes all branches of medicine, psychiatry, surgery, and related disciplines. Clinical research includes all forms of research, including questionnaires and survey studies; noninvasive and invasive screening, diagnostic, and therapeutic procedures; genetic studies; and future studies with not-yet-formed scientific hypotheses involving stored data and biological specimens.

The precision required of informed consent in clinical research has not been matched in the realm of clinical care. Principal investigators often view the disclosure requirements of informed consent in research as stumbling blocks, especially if they are more familiar with the concept of informed consent as it is used in clinical care. The IRB is required to apply a much higher level of scrutiny to informed consent for research than has been applied, in general, to informed consent for clinical care.

First, let's look at the history of informed consent as applied to clinical care.

Informed Consent in Clinical Care

In the clinical practice of medicine in the United States today, it is recognized that, from the judicial perspective, there are two main standards of informed consent: the "professional standard" and the "reasonable person standard." Under the professional standard, a physician must inform a patient to the extent that other physicians in his or her "community-of-peers" would inform their patients. Under the reasonable person standard, a physician must inform a patient to the extent that a reasonable person in that patient's position would wish to be informed. However, ethicists have emphasized a third standard, that of the subjective patient, which takes into account unique perspectives of the individual patient.

In the history of informed consent in clinical care, there have been four landmark court cases:

1767, *Slater v. Baker and Stapleton* (Great Britain)
1914, *Schloendorff v. Society of New York Hospital* (New York)
1957, *Salgo v. Leland Stanford Junior University Board of Trustees* (California)
1972, *Canterbury v. Spence* (District of Columbia)

Consent, 1767 to 1957

The first key concept of consent appeared in a court case in 1767 in Great Britain, *Slater v. Baker and Stapleton.*[1] The court did not create a standard of consent but, rather, took testimony from physicians regarding whether the obtaining of consent from a patient was a custom of the profession. This was an adoption of the professional standard of disclosure.

Self-Determination, 1914

In 1914 in the United States, Judge Benjamin Cardozo argued that the basis of a patient's right to consent to treatment was the patient's right of self-determination.[2] Judge Cardozo took *self-determination* as a basic right, yet there was a provision in his statement that self-determination was a right of a patient *of a sound mind.*

Informed Consent, 1957

The term *informed consent* did not enter the judicial lexicon until 1957, when it was used by the judge in *Salgo v. Leland Stanford Junior University Board of Trustees.*[3] In *Salgo,* there was no elaboration of the concept of informed consent, and the judicial opinion was criticized for introducing the term *informed consent* without explaining how it was to be used in judicial decision making.

The Professional Standard, 1767 to Present

Between 1767 and 1957, there was only one disclosure standard: the professional standard. Under the professional standard, a physician had to disclose to a patient whatever information the physician's peers in good standing would disclose to their patients. The courts heard testimony about what physicians said similarly trained physicians would disclose, but it was not clear how such agreement was reached among physicians.

From 1767 to 1957, courts simply assumed that there was a professional custom of securing a patient's consent to treatment. There were many questions that went unasked: How did physicians come to agree that consent was to be obtained from a patient? How did physicians come to agree on what was to be said in the consent session? How did one ascertain whether a particular patient's consent had actually been obtained?

When the term *informed consent* was used for the first time in the *Salgo* decision, the notion of consent became tied to the concept of information in terms of the nature and content of risk the physician disclosed to the patient. There was still little understanding of how physicians—as a profession—came to an agreement on what information should be disclosed.

The Reasonable Person Standard, 1972 to Present

In 1972, Judge Spottswood Robinson created the reasonable person standard, in the landmark federal decision in the District of Columbia, *Canterbury v. Spence.*[4] Judge Robinson argued that there never really was a professional standard of disclosure in medicine. But even had there been, a standard of information disclosure based on what physicians thought was important would not be the appropriate judicial standard. Judge Robinson argued for the reasonable person standard, in which a physician was to inform a patient of what *the reasonable person in that patient's position* would want to know.

Today, in judicial situations, two standards—the professional standard and the reasonable person standard—are dominant in the United States. The professional standard holds a slight prevalence over the reasonable person standard in adoption by courts in the United States: roughly 52 percent of states hold to a professional standard of disclosure and 48 percent hold to a reasonable person standard. Interestingly, Great Britain still holds to the professional standard only, and the Canadian Supreme Court adopted the reasonable person standard of Judge Robinson's decision in *Canterbury.*

The Patient's Right to Information and "Self-Decision"

Judge Robinson, in *Canterbury,* held that the physician had an obligation to disclose information to the patient and that the competent adult patient had the right of "self-decision" (basically, that every adult of sound mind has the right to determine what shall be done with respect to his or her own body).

Pointing out the informational needs of patients, Judge Robinson declared that competent adult patients have a right—after receiving information from the physician about the recommended medical intervention, its risks, and its alternatives—to base a decision on whatever grounds they see fit, be they rational or irrational in someone else's view.

Judge Robinson set three types of information as being important in informed consent: information about the nature of the procedure, its risks, and its alternatives. Yet, he emphasized risk disclosure, for the important reason that this was the issue that patients were bringing to courts for adjudication. Patients were incurring risks related to medical interventions recommended by their physicians, and the patients alleged that they had not been informed about these risks. This is why most of the informed consent cases were brought before courts. Other issues of informedness have, of course, been brought to courts for adjudication, such as a patient's alleged lack of information about the scope of an intervention in a surgical procedure as contrasted to what the patient understood regarding what was to be performed in the operation. But risk disclosure has been the point highlighted in most judicial decision making related to informed consent in clinical care.

The Subjective Patient Standard

A third standard, the subjective patient standard, has been developed by ethicists as an optimal standard for informed consent. This standard focuses on the unique patient, with his or her own values and life goals, who is considering a set of alternative medical interventions, including nonintervention (letting Nature take its course) and delayed intervention. It is primarily an ethical standard; the courts, up to the present, have not seriously considered the subjective patient standard as a viable standard in judicial deliberations about informed consent. It received particular attention from Ruth R. Faden and Tom L. Beauchamp in their 1986 book *A History and Theory of Informed Consent.*[5]

Prospection and Hindsight

The key points in most judicial deliberations about informed consent relate to the issues of "hindsight bias" and "retrospective versus prospective decision making." In most court cases regarding informed consent, the patient is alleging insufficiently informed consent after the fact of an injury. Thus, the patient's reflections on what he or she might have wanted to know before the procedure are complicated by the fact that an injury occurred. Judge Robinson recognized this "hindsight bias" in the *Canterbury* decision and argued for the reasonable person standard over the subjective patient standard in judicial decision making. A reasonable person, he said, would want a disclosure of information regarding risks before making the decision regarding an intervention the physician was recommending. The disclosure of risks prospectively (before a medical intervention) was intended to eliminate bias when looking back on the decision after the patient had sustained an injury from the intervention.

This concept applies directly to informed consent in research. Because study participants must give their informed consent prospectively, based on what ideally will in hindsight seem to have been adequate information, the IRB is involved in the informed consent process from the start, to ensure that the principal investigator will present individuals the best and complete information for prospective decision making. This is why IRBs spend an enormous amount of time making certain that all disclosures are being made at the time the individual is considering whether to participate in a research study. The requirement of written consent and the insistence that a person take the consent form home before signing it allow for additional reflection on what was discussed in the informed consent session and for discussion of the decisions with families or significant others. After further reflection and discussion, a potential participant may decide not to enroll in a research study. Informed consent in research allows the participant the right of terminating study participation at any time when it can be done so safely.

Informed Consent and Decision-Making Capacity

The courts have usually assumed the patient to be competent to grant consent. The issues of competency to consent to treatment are clearly related to those of decision-making capacity. Paul S. Appelbaum and other researchers consider such competency to include four abilities: to understand, to appreciate, to reason, and to evidence a choice.[6] Regarding informed consent, the basic issues of cognitive capacity to make a decision (decisional capacity) include an understanding of disclosed information, an appreciation of the state of affairs surrounding one's own clinical state, working with information to come to a decision, and evidencing a choice.

A person can be said to understand the information contained in an informed consent form if he or she can repeat the information and/or can answer a set of true-or-false questions regarding the information. Although such individuals can be said to have an understanding—in some sense of the word—of the information disclosed in the informed consent form, some of them may not be able sufficiently to appreciate the ramifications of treatment or may lack an appreciation of their own clinical state. Let us take the case of someone with severe acute mania.

A person in a state of severe acute mania may be able to repeat information disclosed in an informed consent form and may be able to answer correctly a set of true-or-false questions, but that individual may not be able to recognize that he or she is in need of treatment. Thus, this patient would not be considered to have sufficient decisional capacity to give consent because of an inability to appreciate that he or she is in a state of severe acute mania.

The phrase "working with information" refers to the cognitive operations an individual performs in structuring and weighing information to prioritize it in the task of coming to a decision. In this context, the structuring and weighing of information relates to the notion of considering risk, weighing risk and benefit, and considering personal values in regard to the willingness to accept risks of a medical intervention and to proceed with that intervention.

In "evidencing a choice," the individual must be able to take the information and prioritize it to come to a decision. A number of factors can prevent an individual from being able to evidence a choice (see Box 10.1).

Box 10.1. Factors Inhibiting Decision Making

- The individual is presented with less information than he or she believes is needed to come to a decision.
- The individual is confused by the information.
- The individual is not able to keep track of all of the information.
- The individual is not able to decide which information deserves more weight according to his or her personal values.
- The individual recognizes that he or she cannot make a decision unaided and needs a significant other to become involved in the decision.

In clinical research, two of the challenges facing all principal investigators are developing measurement tools to assess decision-making capacity in a scientifically valid way and developing the clinical criteria upon which to assess an individual's capacity to consent to participate in research.

The Belmont Report and the Protection of Human Research Participants

The National Commission for the Protection of Human Subjects of Biomedical and Behavioral Research came into existence when the National Research Act (Public Law 93-348) was signed into law in 1974. The commission was charged to identify basic ethical principles and to develop guidelines that clinical researchers would follow.[7] The commission met for an intensive four-day period of discussions in February 1976 at the Smithsonian Institution's Belmont Conference Center. The commission's report, published in 1978, has therefore come to be known as the Belmont Report. It is a statement of basic ethical principles and guidelines to assist in resolving ethical problems in the conduct of research with human subjects.

The Belmont Report rejected the professional standard in clinical research, arguing that the foundation of the professional standard was that some type of common understanding existed among physicians on what was to be done medically and what was to be said to patients. However, the commission said, the professional-based approach is inadequate for clinical research because "research takes place precisely when a common understanding does not exist." The Belmont Report also rejected the reasonable person standard, which requires that the information communicated to a patient be "the information that reasonable persons would wish to know in order to make a decision regarding their care." Applied to clinical research, this standarad would refer to the information desired for an individual's decision to participate in a particular research study. The commission argued that the reasonable person approach is insufficient, because "the research subject, being in essence a volunteer, may wish to know considerably more about risks gratuitously undertaken than do patients who deliver themselves into the hand of a clinician for needed care."[8]

Instead of accepting either of the two predominant judicial standards of informed consent in clinical care, the commission created a new standard, which it called the standard of *the reasonable volunteer.* The Belmont Report describes this reasonable volunteer standard as follows: "the extent and nature of information should be such that persons, knowing that the procedure is neither necessary for their care nor perhaps fully understood, can decide whether they wish to participate in the furthering of knowledge." The report adds the following point: "Even when some direct benefit to them is anticipated, the subjects should understand clearly the range of risk and the voluntary nature of participation."[9]

Thus, the standard for informed consent delineated by the National Commission for the Protection of Human Subjects says that individuals being recruited into research studies need to understand clearly the range of risks they face and the fact that they are *voluntarily* participating in the research study.

11 Continuing Review, Communication, and Feedback

Continuing Review

The IRB's initial consideration of a research study for approval or rejection, its discussion of the reasons behind the decision, and the reevaluation of the study after revisions are only the front end of the IRB's work. The institutional review board must also address continuous quality improvement.

When (if) the IRB approves a study with human participants and the medical institution allows the project to proceed, the IRB begins an ongoing supervision and periodic review of that study. Continuing review includes many components that the IRB must become acquainted with: thorough review of the study, any complaints, any adverse outcomes that occur to participants, and much more. The IRB monitors participant safety during the entire research project and even beyond the termination date, if unanticipated results occur. The IRB's supervision continues as data are analyzed and findings are discovered.

Continuing oversight of an approved study requires that the IRB systematically review all key issues listed in Box 11.1.

Box 11.1. Areas Subject to Continuing Review by the IRB

- the scientific protocol
- the informed consent form
- reports of adverse outcomes
- reports from other monitoring boards
- unanticipated problems
- any noncompliance with study parameters
- the study findings to date
- any updates in recruitment materials and study brochures

The Frequency of Continuing Review

The IRB is required to review the entire research study within 364 days of the meeting at which it approved the study. However, for higher-risk studies, reviews might have to occur more frequently. The IRB decides how often it should review each study—perhaps after the first participant is enrolled; after the first three participants are enrolled, in the first thirty days, the first quarter, every quarter, or only annually. Continuing review of studies at appropriate intervals and responding promptly if problems arise are key to minimizing adverse outcomes.

Tracking Studies

Prioritizing Information for Review. The IRB must prioritize the information that comes into its office so that the information most pertinent to the protection of study participants is always up front and ready for the IRB to consider and act on. Risk information and summary information on study findings to date rank highest, but there are other key issues to be addressed in a continuing review (see Box 11.2).

Box 11.2. Some Elements of a Continuing Review

- Does the study involve an investigational new drug or device? If so, there may be a mandate for a full review of the study at specified intervals.
- Is the research permanently closed to the enrollment of new participants? If not, has enrollment of additional participants raised new issues?
- Have all participants completed all research-related interventions, and will the research remain active only for long-term follow-up of participants?
- Have any additional risks been identified?
- Have new relevant data or risks been revealed in the peer-reviewed medical literature? Have problems occurred at another site of the study?
- Has the study reached the stage where all remaining research activities are limited to data analysis?

Once a study is under way, the IRB must keep track of various types of reports:

- Peer-reviewed medical journals will publish reports and papers about a study's progress at different sites. An up-to-date IRB member may know about findings before the principal investigator or study sponsor sends reports of those findings to the IRB.
- Through the principal investigator, the study sponsor will send the IRB reports of any illness or deaths occurring in study participants at other sites (reported to the off-site IRB by data-monitoring boards).

- Principal investigators will report any on-site illness or death.
- Study participants may make complaints.
- Reports of a principal investigator's or research staff's potential noncompliance with IRB recommendations may come in, requiring investigation.

Tracking Where a Study Is at a Particular Time. The IRB should always know the stage a study is in at a particular time—not started, beginning phases, middle phases, ending phases, closed to participant enrollment.

Once a study is closed to participant enrollment, it will be in one of the following stages: data being analyzed for interpretation, data being reanalyzed and interpreted, data being written up in a report or scientific article for consideration of publication in a peer-reviewed medical journal.

Even in the data-analysis phases, there may be information regarding results that will have to be communicated to the participants, and the IRB will need to know about this. An example of such information would be new beneficial data that have accrued or new adverse data that were not apparent in an earlier phase of analysis of data.

Review of Documents and Reports of Adverse Events

The IRB has to address the scientific and ethical issues of any amendments to the study's original documents. Amendments and modifications to the scientific protocol must be considered for approval, rejection, or modification and reconsideration. Once an informed consent form is approved by an IRB, that form cannot be changed in any way without prior approval of the IRB.

Adverse events reported both on-site and from other sites in which the research study is being carried out must be analyzed. If new adverse outcomes of an unexpected and severe nature are reported, this may necessitate a change in how the entire research project is being reviewed and interpreted and may affect whether the study can continue as designed.

The IRB should educate principal investigators and research staff members in the uniform and complete reporting of adverse outcomes. Thorough reporting is essential for IRB members to understand the significance of the adverse outcome in relation to the research study and to act appropriately and promptly for the benefit of participants. Basic questions regarding adverse outcomes are shown in Box 11.3.

Box 11.3. Questions about an Adverse Outcome

- What is the precise nature of the adverse outcome?
- What is the level of severity of the adverse outcome in the particular participant's case?
- What is the estimated etiology of the adverse outcome?
- Could the adverse outcome have been associated with participation in the study?
- Could the adverse outcome have been associated with negligence on the part of the research staff? For example:
  —Was the adverse outcome due to lack of oversight by the research staff of a research-related laboratory?
  —Was the adverse outcome due to lack of action on a study finding?
  —Was the adverse outcome due to delay in action on the part of the research staff?
  —Was such lack of action due to failure to notify the participant or the participant's health care provider?
  —Was such lack of action due to poor communication of an abnormal laboratory result or study finding?
- Has the principal investigator modified the informed consent form to include in clear and precise lay language the possibility of the adverse outcome's occurring in future participants?
- Has the principal investigator promptly given the modified informed consent form to the IRB for review, consideration, and approval or modification?
- Has the principal investigator notified all research staff members about the adverse outcome in sufficient detail and with a clear plan of how the research staff is to be on alert for this problem in the future?

The IRB may receive reports from the principal investigator regarding adverse outcomes that may be related to the study drug or device. IRB members must decide if they agree with the principal investigator's and study sponsor's determinations regarding the cause of the adverse outcome. Do they agree on whether the adverse outcome is *unrelated* or is *possibly, probably,* or *definitely* related to the drug or device? This rate of chance may be described in numbers or in words or both.

Especially with a multisite study, the new IRB member must quickly adjust to the ways in which adverse outcomes are reported. In a multisite study, the

Box 11.4. The Adverse Outcome Report

- Each report of an adverse outcome must be accompanied by an estimate of whether the adverse outcome is related to the drug, device, or intervention being studied, which may be described as
  —not related,
  —possibly related,
  —probably related, or
  —definitely related.
- The IRB must evaluate each report of any new adverse outcome as a new risk of the study drug or device.
- Upon report of a new adverse outcome, the principal investigator must begin notifying participants enrolled in the study and participants who have left the study about the newly reported risk of the study drug or device.
- The IRB must evaluate each unanticipated adverse outcome for possible relevance to the risk section of the informed consent form, which may need to be modified in light of the outcome.
- Most important, the IRB must decide if the study
  —may continue,
  —may continue with modifications,
  —must be suspended, or
  —must be terminated.

new IRB member will need to adjust to how other sites formulate their reports of adverse outcomes. IRB members get used to the way their own board formulates and reports adverse outcomes at its own site, and an adverse outcome report related to a new medical product may be quite complex. In reviewing off-site reports, the local principal investigator's input should be sought.

Tracking Reports of Adverse Outcomes

The collective memory of an IRB records how decisions were made at past meetings. Collective memory can be helpful because IRBs have easy access to it during meetings, but it may be problematic if the IRB members' memories differ on the specifics of the discussions. Collective memory is no substitute for accurate meeting minutes. However, if meeting minutes are not detailed, they may add nothing to the collective memory.

Even the strongest collective memory of the IRB must not be relied upon to track reported adverse outcomes. The IRB should have a mechanism in place to track adverse outcome reports over time, to see, among other facts, how often they are occurring. The IRB has the responsibility to notify the principal investigator if too many adverse outcomes are occurring. If they are, the study should be stopped until the principal investigator, product manufacturer, or Food and Drug Administration provides more definitive answers regarding the adverse outcomes.

The institution's research service can help develop a program for tracking adverse outcomes. The program should record all adverse events that principal investigators report to the IRB, specifying the number, type, and degree of severity of adverse outcomes occurring at the IRB's site and at other sites in the study. This program will give the IRB a perspective on each study it oversees. The information discovered by means of the tracking program should then be compared to the risks stated in the informed consent form, to make certain that the form is continually updated with respect to all adverse outcomes reported in a study.

The IRB must develop interoffice frameworks that allow the recording of data in a format that facilitates the identification of trends, so that the IRB can determine whether the research study and its informed consent form need modification or revision or the research must be suspended or terminated. The more quickly the IRB can gain access to cumulative data and data trends in particular studies, the more time the members will have for deliberation. It is appropriate for the medical institution to make a commitment to the IRB in staff and personnel, easy-to-use databases, and statisticians to help analyze cumulative data and trends in data over time.

Risks related to a study can be reported to the IRB and captured in a database as falling into the categories listed in Box 11.4: not study-related, possibly study-related, probably study-related, or definitely study-related. For tracking purposes, it is important to include in the database: the key descriptors of what happened to the participant who sustained the adverse outcome, all reports to the IRB of the opinions of others (and who the others reporting opinions were), and the opinions of the IRB members when the issue is discussed by the full board.

IRB Action after Adverse Outcomes

When adverse outcomes occur during a research project, the IRB must decide whether to allow the study to continue, to stop it until the principal investigator and IRB can modify the study appropriately, or to end the study. The IRB has, as stated by the *Code of Federal Regulations*, the "authority [and obligation] to suspend or terminate approval of research that is not being conducted in accordance with the IRB's requirements or that has been associated with unexpected serious harm to subjects."[1]

Anticipated adverse outcomes have already been identified in the informed consent form a research volunteer has studied and signed. Unanticipated adverse outcomes were not disclosed in the informed consent form, although the possibility of them should have been. The IRB must decide whether all participants in the study should be apprised (e.g., by a letter sent by the principal investigator and/or product manufacturer) of the new adverse outcome. The board must also decide whether the scientific protocol and informed consent form that are in current use should be amended, and presented to participants, explaining the risks that are unfolding regarding a study drug or device. The investigator, study sponsor, and IRB must sort out whether this new, unanticipated risk should in fact have been anticipated and specified in the risk section of the original informed consent form. However, they need quickly to decide if a revised informed consent form is needed, because the volunteers enrolled in the study may decide to terminate their enrollment after hearing about the newly identified risk.

Continuous Institutionwide Surveillance of Compliance with Regulations

Although the IRB's primary focus in all of its work is the protection of study participants, the entire medical institution must provide continuous surveillance to ensure that principal investigators and research teams comply with research regulations. Those who must remain informed include the institution's director, his or her office personnel, the chief of staff and associated personnel, the research director and the research office staff, and all IRB members and IRB coordinators.

Any medical institution that is involved in both the care of patients and research on human participants can be considered to have both clinical and research components. Yet, within a particular institution, the clinical care of patients and research on human participants occur in the same outpatient clinics and medical, surgical, and psychiatric departments. Thus, in all sectors of the medical institution, health care providers caring for patients are also caring for participants in research studies. The clinical staffs of medical institutions are well positioned to help detect and report possible irregularities in research that the medical institution is required to attend to. Processes must be in place to immediately address, evaluate, and, if necessary, remedy the irregularity, for the protection of patients and study participants. In the protection of study participants, both sides of the hospital (the clinical side and the research side) must have continuous knowledge and understanding of research that is going on there and the regulations under which it is being conducted.

Ideally, no research project should take place in a hospital that the clinicians are not aware of, and there must be clear and precise communication between the medical institution's research personnel and clinical personnel. This is particularly true when research would involve the use of new medical procedures or interventions. The clinical service, research services, and IRB may benefit from one another's expertise in planning how to protect patients and human participants as any new procedure is developed in the medical institution. In the case of new procedures and interventions, the protection of patients and the protection of study participants overlap, and all expertise of the medical institution must be brought to bear in a coordinated way.

Ongoing Communication and Feedback

Communication and feedback from principal investigators, research staffs, study participants, and experts is essential for the best functioning of an IRB. The IRB must have and keep open channels of communication with all parties involved in studies and everyone within its own institution, institutions with which it has dealings regarding the protection of study participants and ongoing or future research, and regulatory bodies that it interacts with to make certain it is proceeding correctly along the course of optimal protection of participants.

The goals of communication in the context of research on human participants are to help all parties understand and execute their role in protecting participants. IRB members are role models within medical institutions for their focus on the protection of participants and are resource people for how research on human participants can be continually improved.

There must also be easy and full communication among the members of the IRB. The board's best work results from the combination of all the members' knowledge and opinions.

Communication between the IRB and the Principal Investigator

An IRB must maintain open lines of communication with all principal investigators active at its institution. There is much communication between the two parties during the initial approval process. If a principal investigator's scientific protocol or informed consent form contains ambiguities, the IRB can ask the principal investigator to answer questions and make modifications to improve the clarity. The IRB can invite the principal investigator to further elaborate on the scientific protocol and informed consent form in a meeting of the full board. During the study, the IRB should solicit feedback from principal investigators regarding how participants are reacting to the recruitment process and to study participation, and particularly about any problems that were not predicted by the principal investigator or the IRB. The principal investigator may seek contact with the IRB at any time a question arises and should do so if he or she notes that individuals are having trouble understanding the informed consent form.

Communication between the IRB and the Research Team

Although the IRB deals directly with the principal investigator, the IRB's support staff will often communicate with someone on the research staff (for example, the study coordinator) to see that things are attended to in a timely fashion and to ensure that there is smooth, effective communication between the IRB and the principal investigator. The IRB sees to it that the entire research team is educated in the ways to optimally protect study participants, to allow the safe conduct of the study. This training includes effective, complete, and honest communication of information to the IRB at all times.

Communication between the IRB and the Study Sponsor

Most of the IRB's interaction with a study sponsor is through the principal investigator. However, when, in the IRB's opinion, the study sponsor (e.g., a product manufacturer) is not optimally protecting human participants in the design of the scientific protocol or in statements in the informed consent form, the IRB must make certain that the principal investigator communicates that message to the study sponsor so that the IRB's instructions regarding the protection of participants can be followed. If it appears that the IRB's messages to the sponsor are not being attended to properly, it may communicate directly with the sponsor.

Communication between the IRB and the Study Participants

The IRB should designate a member (e.g., the chair of the IRB) whom participants may contact regarding any issues that arise during the conduct of the study. Regardless of who is chosen as liaison with study subjects, the IRB chair's name and telephone number should appear on every informed consent form approved by the IRB. In addition, participants must be given contact numbers for the principal investigator, so they can discuss issues and ask questions regarding participation in the study.

The IRB needs to receive feedback from participants regarding the conduct of the study, the actions of the research team, the quality of the informed consent form, how well the institution is performing in the way the research is being conducted, and any suggestions on how the IRB and the research team could do a better job.

Conversely, the IRB will need to contact participants under various circumstances, for example, to inform them of unanticipated adverse events.

Communication between the IRB and Health Care Providers

The IRB needs open and full communication with the participants' own clinical caregivers and with the health care providers of the medical center(s) where the study is being conducted. Often, health care providers not directly involved with a research project will offer insights into what is going right and what is going wrong with the study, from their vantage point in the clinical environment. They may have observations on the way in which individuals are being recruited into the study, the way in which primary care providers are being notified of their patients' participation in the study, and the way in which adverse outcomes are being handled if a participant calls his primary care practitioner instead of the principal investigator or research team contact.

Health care providers whose patients are study subjects need easy access to the project's scientific protocols and informed consent forms, to know whether a symptom a patient develops could be related to the study. For example, if, one year after participating in a drug study, an individual shows a loss in white cells, red cells, and platelets in the blood, could this represent a delayed impact of the study drug on the bone marrow? Secure electronic medical record keeping allows some institutions to provide a pop-up screen each time a health care provider enters a medical record to show what research studies an individual is or was participating in. Such electronic medical records systems can be expanded to contain not just a listing of past or present studies but also copies of the scientific protocol and informed consent form and contact numbers for the principal investigator, the IRB, and the research service so the health care provider can obtain relevant information about the risks of all study drugs the participant has taken.

Communication between the IRB and Its Parent Institution

The IRB must be able to interact with all departments of the institution where the research is being conducted, to assure that study participants are protected across the various medical services. For example, often, a medical center employee or a patient will see and bring to the IRB's attention a recruitment advertisement that he or she considers either coercive or suggestive of influence on the part of the principal investigator (e.g., offering an intervention through the study that is currently being restricted in patient care at the medical institution because of cost constraints).

IRB minutes are reviewed by the research department or the research and development committee of the institution and by the boards of directors and legal officers, and are signed by institutional leaders. This review structure provides opportunities for dialogue between the IRB and the staff of their institution. The IRB should also contact the leadership of the institution any time they feel that there is a problem with respect to the protection of human subjects within the institution or any similar issue.

Communication among the IRB, the Principal Investigator, Other Services, and Other Institutions

The IRB can often connect the principal investigator with the departmental chiefs and service chiefs whose involvement will be needed to ensure that the study participant is protected with the best expertise in the medical institution.

Continuous management is required if the study topic has potential impact outside the principal investigator's area of expertise (for example, a surgeon studying aspects of diabetes and vascular disease will need the oversight of an endocrinologist).

Often, the IRB can help the principal investigator contact local experts who have successfully protected participants on scientific protocols with similar study designs. For example, a diabetes specialist can be contacted to determine whether the range of hemoglobin a1c values being assessed in a sponsor-developed protocol is acceptable according to optimal medical care.

In multisite studies, IRBs receive reports concerning adverse outcomes from other institutions and from data-monitoring committees. Through these reports, IRBs see how aspects of reporting are being handled at other institutions and can discuss with other IRBs why differences occur, what to do about adverse outcomes that are being treated differently and how each views adverse outcomes.

Another example of an opportunity to share experiences across IRBs is when an IRB member serves on more than one institution's IRB. This person can discuss with each board the areas of similar function and of divergence in the treatment of key issues.

Finally, IRBs can directly contact each other to discuss challenges they face in devising programs to protect study participants and train themselves and research teams.

Communication between the IRB and Regulatory Bodies

When questions or problems arise or when issues come up that might be clarified by discussion with regulators, the IRB should communicate directly with regulatory bodies, such as the FDA, to learn how regulations are interpreted by those regulators and how best to abide by the regulations.

There are occasions when the IRB may need to contact specific regulators, for example, if a new medical device is being studied within an institution and the IRB has a question regarding the device's status with the FDA in terms of approval. The IRB will need to contact the specific individuals within the FDA who will be most familiar with the device, its regulation, its status, the risks as viewed by the FDA, and other issues.

Soliciting Feedback

The IRB should solicit feedback from all parties involved in a study, particularly within its institution. Anonymous suggestion boxes can be used to initiate discussions of how to optimize the protection of participants. The IRB chair is often approached by clinicians and researchers who want to comment anonymously on particular IRB decisions or the actions of research teams or principal investigators, and the IRB chair can bring these comments to the board while preserving those persons' anonymity. The institution's research service should have in place a formal mechanism by which anyone can raise issues for consideration, and there should be an environment of active discussion throughout the institution on matters of how best to protect participants. Such opportunities for communication allow research to continue and be fostered not only by researchers but also by the public.

Education through Communication

One of the strongest ways an IRB has to educate is through its communications with other parties in research projects. The principal investigator obtains a theoretical perspective on the protection of human participants while sitting in a classroom, viewing a training video, viewing slides on a computer, or accessing a website focused on how to best protect human participants. The principal investigator obtains a real-world perspective when he or she sees the IRB's response to his or her scientific protocol and informed consent form. The IRB applies the theory of the classroom to the reality of these documents.

There is often a distance between the interests of the principal investigator and those of regulatory bodies. The IRB can help the investigator interpret the regulations in the conduct of research studies. In addition, the IRB learns from relevant experts who are brought into the discussion to share opinions or help settle disagreements regarding the application of regulations to specific scientific protocols.

Federal Regulatory Bodies

Another party in the communication and continuing review processes is the federal regulators who oversee research on human subjects. They can sometimes facilitate communication among the other parties, and they are charged with making periodic reviews of the work of IRBs.

When regulators come to a medical institution to review its IRB, the IRB's meeting minutes are crucial in documenting the range of discussion and the differences of opinion within an IRB on a particular topic. Meeting minutes should always detail the full discussion of the IRB, covering all opinions on all topics.

Differences of opinion among IRB members may include how federal regulations should be interpreted. Regulators can guide the IRB in the interpretation of federal regulations and should be consulted when such issues do not resolve easily. Differences of opinion among IRB members may extend to the very approach of the IRB to the best protection of participants. Let us take the example of an IRB on which some members disagree and other members agree with the principal investigator that certain risks should not be disclosed or that there is no need to develop a better instrument to measure participants' decision-making capacity and capacity to consent to research. Regulators should be consulted for their recommendations regarding how decision-making capacity should be assessed. These issues are difficult for all parties. New ideas may come from a dialogue between the IRB and federal regulators.

A complicated and sensitive problem that may inhibit communication and the IRB's decision making is conflict of interest within the IRB. Some IRB members may feel, for instance, that one or two members are protecting research agendas within their own division of the institution, and these concerned members may feel that they need a new forum for discussion and new input into the debate. Discussion of the issue with regulators provides an external voice and may produce new insights among IRB members. However, in such a situation, the IRB members must obtain assurrance from the regulators that the members' views will be held in confidence. Often, the regulators do not discuss what information they will keep confidential, and an IRB member may be concerned for his or her job. Without clear specification of what is confidential information, an IRB member may hesitate to speak.

In all discussions and communications regarding any sensitive topic within a medical institution, issues of freedom of speech and confidentiality must always be clarified. The United States is fortunate to have IRBs at the local level, involving local personnel in decision making rather than having research proposals approved or disapproved by a distant authority. Yet, if an IRB member lacks confidence in speaking up, the foundation of the IRB's decision making will be eroded and its ability to optimally protect study participants damaged.

12 Where Are IRBs Making Mistakes, and How Can We Minimize Mistakes?

IRBs work under conditions of insufficient resources and time for the performance of their numerous duties. In these circumstances, IRBs can and do make mistakes when they review and deliberate on research and make recommendations. As a consequence of IRB mistakes, research programs at several major research institutions in the United States have been closed for periods of time. These shutdowns were to allow the IRBs to examine their practices, reflect on the *Code of Federal Regulations,* and restructure the way in which they review research. The materials that have been published about these suspensions highlight certain themes regarding how IRBs may deviate from their main task of protecting the human participants of research.

Mistakes at the level of the IRB can come from many sources, and the IRB must be cognizant of the potential sources of these mistakes so that it can recognize mistakes early and avoid repeating them.

Deviations from the *Code of Federal Regulations*

The first potential source of IRB mistakes is a failure to follow the *Code of Federal Regulations (CFR).* Nonadherence by an IRB to any federal regulation is a problem. Nonadherence may come to light only when a regulatory body's team arrives for a site visit. At the time of the site visit, the team may ask why, for example, the IRB is not recusing conflicted members during discussions and votes on research proposals. The IRB may argue that its meeting agendas are so full that it cannot take the time to recuse conflicted members. The institution and the IRB will then have to reexamine how each conducts its business, and the institution must help the IRB obtain the additional time to recuse conflicted members.

The Presence of a Quorum

If the IRB conducts business without a quorum present, it is not following regulations. There can be distortion of an IRB's decision when some members are absent, even when a quorum is present. Say that with all 20 of the members of an IRB present the vote on a proposal was against the study, 12 nay to 8 yea. If 6 members had been absent who would have dissented, the majority would have shifted, 6 nay to 8 yea, and the proposal would have been approved. The distortion is not intentional, because the absent members were not enticed away, yet the vote did not represent the full IRB's decision. Attendance at IRB meetings is essential at all times to ensure that all opinions are heard and that votes taken represent the full board's opinion on a matter.

The Presence of a Nonscientist

The *CFR* specifies that each IRB shall include a member who is a nonscientist. The IRB's nonscientist must be present during the meeting, especially during deliberation and voting. If the nonscientist is absent, the discussion and vote must be postponed until the nonscientist is present.

Conflicts of Interest

The IRB can fail to note that the principal investigator and study sponsor did not specify the conflicts of interest that are inherent in a research study conducted by a specific medical research institution, by a specific principal investigator, and reviewed by an IRB with a specific membership. There can be a broad range of conflicts that need to be considered for full disclosure. For example, the study sponsor may have endowed a professorship in the medical research institution with the stipulation that the position's occupant conduct studies on its product. Or the medical research institution, the study sponsor, and the principal investigator may all be involved in ownership of patents based on research derived from human cells and tissues collected from study participants. Or the principal investigator or IRB member may own shares of stock in the study sponsor's company. Some conflicts of interest will not affect the validity of a study, but all need to be noted.

Interpreting Regulations

An IRB can have a problem interpreting a specific regulation yet forge ahead without seeking clarification from the appropriate regulatory body. For example, a principal investigator can send a letter accompanying a research proposal asking that the IRB approve the study as a minimal risk study. The "minimal risk" designation should be made only after careful consideration. When IRB members are in doubt about the application of a concept in the federal regulations, the IRB should always seek guidance from federal regulators.

Failure to Perform a Complete and Thorough Review

Members Not Admitting to Not Understanding the Study

IRB members should never be afraid to say that they do not understand what is going on with a study or that they need further clarification. It takes time to develop the ability to accomplish a penetrating review of a scientific protocol. A new IRB member should never hesitate to ask a question, because that question may open the eyes of the other IRB members to an issue that they had failed to take into account.

Not Enlisting Expert Opinion

An IRB must have a lot of experience to understand when it does not have enough information to properly address a scientific study. When there is need, an IRB must call in local, regional, or national experts for help in understanding an aspect of a scientific protocol, a research method not familiar to the board members, or an intervention that is not part of standard care. Expert help beyond federal regulators, such as legal specialists, may sometimes be needed in interpreting guidelines and regulations.

Not Correctly Assessing the Level of Risk

Sometimes an IRB does not accurately gauge the level of risk entailed in a scientific protocol. Perhaps the IRB does not understand the science well enough or the principal investigator has not adequately clarified the science. A principal investigator may frame a scientific hypothesis to appear less risky than an independent review of the scientific hypothesis would suggest. At minimum, the IRB must thoroughly search the peer-reviewed medical and scientific literatures for any problems that occurred in earlier studies. Although systematic review of the literature should be the job of the principal investigator and study sponsor, the IRB needs to double-check if there are potential problems that the principal investigator is not explicitly recognizing at the time of the submission.

Not Searching the Peer-Reviewed Literature Thoroughly

IRBs have relied perhaps too heavily on materials provided by the principal investigator and have not spent enough time searching the peer-reviewed medical literature on the science and the known risks in the scientific protocol. The primary responsibility for checking the validity and correctness of all submitted materials rests with the principal investigator and study sponsor, and this responsibility includes systematic searching of the peer-reviewed medical literature. However, the IRB must conduct an independent search of the literature to verify that the principal investigator and study sponsor have reviewed the literature thoroughly. Unless it double-checks the literature (e.g., for reports of risks of study drugs or interventions), the IRB cannot offer any assurance that the principal investigator's and study sponsor's searches were exhaustive. The IRB's check can be educational for principal investigators and study sponsors, and it provides important documentation for the IRB regarding the quality, good or bad, of the research behind the proposal.

The IRB must determine how it will accomplish the necessary literature searches. For example, the IRB could assign one member the task of conducting searches during the meeting as questions are raised. An astute member familiar with PubMed and search terms, and a fast typist, may be able to accomplish on-the-spot searches of questions like "Has this particular hypothesis ever been studied in the past?" Other members may be assigned questions to research before the next meeting. If the task is too large for the IRB to conduct on its own, it must seek assistance from other parts of the institution.

Not Using Appropriate Exclusion Criteria

An important way an IRB can promote the protection of study subjects is by seeing that the exclusion criteria applied to potential participants are extensive enough to exclude people with identifiable extra risks of harm from the study. The IRB must make certain that all appropriate measurements will be taken before enrolling a participant in the study. If participants will be exposed to a substance, it is crucial for the principal investigator to know (for example, by checking blood levels of the substance if it is measurable in the blood) to what extent the individual has been exposed to the substance in the past. Indeed, in some studies, any prior exposure to a substance may exclude the individual from study participation, and adding another exposure for an individual already exposed to a toxic substance may be an unacceptable risk.

In reviewing exclusion criteria, general questions IRB members may raise about substances and substance exposure include: Are the substances metabolized by the body? If they are, how are they metabolized, or is their metabolism understood? Do the substances simply exist inertly in the body forever, not causing the individual any problems, or will they cause the participants future problems? Are the substances stored in the body? If so, what problems may the individual develop related to this stored substance if never exposed to the substance again and if reexposed? Asking questions of this type will help the IRB decide whether this exposure-related study should be conducted at all, whether the acquisition of this knowledge for future generations is worth the risks of exposing these individuals.

Not Including All IRB Members in Review and Decision Making

The full IRB is necessary to judge the adequacy of a research proposal. An IRB can make the mistake of allowing one or two members of the IRB's subcommittee to take over the job of reviewing a scientific protocol and informed consent form. This type of subcommittee review lacks the variety of perspectives present in the full membership of the IRB.

It is desirable to have all IRB members present for any moderately important or major decision, so that all can contribute to the decision process. No IRB member should have cause to say that he or she has a problem with a scientific protocol after a review that did not include him or her. The protection of study participants demands input from all IRB members, especially for moderate-risk and high-risk studies. Each IRB member's perspective is important on each topic discussed and voted on.

Failure to Adequately Protect Study Participants

Distortion of or Loss of Proper Focus

An IRB can misjudge the safety of a scientific protocol or informed consent form if it becomes overly involved with facilitating research and thus loses its focus on its primary duty of protecting study participants. A research chief may appoint to the IRB people dedicated to the facilitation of research. This may or may not be a bad thing, depending on the intent of the appointments. A research chief may insert a new IRB member into the board with the intention of changing the IRB's attitude or voting. Again, this may or may not be a bad thing, depending on the intent. Given that studies can be approved by only a quorum, often the change of a small number of members will cause the IRB to shift from a committee dedicated to the protection of research participants to a committee dedicated to facilitating research. The purpose of an IRB is the protection of human subjects. Attempts to move the IRB away from that purpose would need to be considered extremely problematic for study participants within the institution.

Not Ensuring that Individuals Understand Their Level of Risk

Usually, the mistake that produces this problem occurs when the IRB is not sufficiently clear and firm with the principal investigator about the content and wording of the informed consent form. For various reasons, an IRB can have a problem counseling principal investigators and study sponsors on full disclosure of risks to individuals considering study participation. All risks must be disclosed. The IRB will need to discuss what it means to disclose a risk and what level of detail about the risks needs to be provided in an informed consent form. As was discussed in Chapters 5 and 11, the information about a study intervention must include how it will manifest itself in an individual; whether the reaction can be life threatening; whether an adverse outcome would be reversible and, if so, to what extent it is reversible; and if irreversible, what individuals can expect in terms of their future life.

In informed consent sessions, the principal investigator may hear from many prospective participants requesting clarification of the risks associated with study participation. The IRB may need to mandate further revisions to informed consent forms based on new information from consent sessions, reports from other study sites, participants who contact the IRB about ambiguities in the consent form or to complain that the form is overly technical, or from newly published research in the medical or scientific literatures identifying new risks.

The IRB must make certain that individuals understand the level of risk they are being asked to assume. This is especially true in high-risk studies involving drugs, medical devices, and exposure of participants to potentially toxic substances. Such high-risk studies include the study of treatments of oncologic (tumor-related) disease or cardiovascular disease. Although these research studies are considered high risk, the medical conditions and diseases under study carry high risk in our population. It is rare in medical research that high-risk diseases can be best understood with low-risk studies.

Coercion in Recruiting Individuals

There should never be any coercion involved in recruiting individuals to participate in a research study, but attempts to influence a potential recruit's decision can be subtle, and a busy IRB may fail to notice them. Most instances of coercion will occur in inappropriately worded statements in an informed consent form. Most will be reported by individuals being recruited into studies. The IRB member should always be on the alert for any wording that may suggest coercion or undue influence. If wording seems odd to an IRB member, it may be misleading to a study participant. The IRB member should bring such odd wording to the full board so the issue can be addressed and the wording clarified or eliminated.

One way of attempting to persuade someone to participate in a study is to exaggerate the likelihood that he or she will directly benefit from participation. The probability of benefit should be honestly estimated and clearly expressed. In those studies in which there is a chance of a benefit, that chance must not be overstated.

Coercion often includes financial matters. There may be direct financial incentives to participate. More subtly, the coercion may be in what is not mentioned in a consent form, and it is up to the IRB to catch these omissions. For example, the IRB may receive an informed consent form that states that the sponsor will not pay long-term compensation for a severe irreversible adverse outcome that occurs to a study participant. In fact, many sponsors do not offer to pay long-term compensation in the case of a severe irreversible study-related adverse outcome. However, in these cases, individuals being recruited should have this pointed out to them, so that they can knowledgeably determine whether they want to enroll in such a study.

A frequently omitted fact is that a participant always has the right to bring suit against a study sponsor to recover compensation. Rarely does an informed consent form specify this right of participants, and the subject is seldom discussed in IRB meetings. The area of clarification of study participants' rights needs further scientific study. In signing an informed consent form, participants do not sign away any rights.

Not Monitoring Ongoing Test Results

IRBs often err by limiting their focus on test results to the baseline laboratory and other tests done upon entry to the study. Throughout the study, the IRB must systematically review the promptness with which abnormal test results are being identified and then communicated to participants and their health care providers. There should always be prompt identification of abnormalities and prompt discussion of them with participants, whether or not they are related to the study; and the participant should be advised on what can be done to prevent harm to the participant, to reverse the abnormality, or to minimize the damage.

The composition of the research team is another potential area of insufficient IRB attention regarding test results. Broad and deep expertise is needed from the relevant areas of medicine to be responsible for (1) deciding which individuals can safely be included as study participants, (2) making certain that all relevant laboratory and other tests are done from the beginning of the consideration of an individual as a potential participant, and (3) immediately attending to and acting on abnormal laboratory and other test results when they are detected.

Analysis of Mistakes

IRBs should talk about their mistakes. A complete analysis of how a mistake came about, needs to be done. Then, the IRB should discuss how a repeat of the mistake can be avoided. Let us return to the example of the scientific protocol and informed consent form that were approved when too few IRB members were present for the vote. Some people examining this event would criticize the

IRB members who were absent. The argument can be made that the way to prevent this mistake from occurring at a future time is for IRB members to ensure that they will be present for all key deliberations and votes. In any IRB discussion, one person's point of view may be crucial in terms of developing a board perspective on an issue. Any time an IRB member is absent from a meeting or for a specific vote means that this perspective may be lost without opportunity for recovery. Such is the complexity and workload of the IRB member carrying out the task of optimally contributing to the IRB.

13 Strategies for Managing the IRB Workload and Supporting IRB Decision Making

The following strategies may help IRBs manage their workload. The aim of these strategies is not only to save time and effort but also to help the IRB refine the criteria that it uses for making decisions and which it will have to defend when regulators and others review its decision making. The IRB develops its own criteria for evaluating the protection provided for participants making decisions regarding the studies that take place in their institution, and it should strive continually to improve these criteria and other methods and procedures.

Many items on the IRB's agenda will recur with nearly every proposal.

Box 13.1. Persistent Tasks on IRB Agendas

- Attempt to secure the presence of all IRB members, including the non-scientist(s), at each vote.
- Develop areas of expertise among the IRB members.
- Assess participants' decision-making capacity.
- Develop tests of decision-making capacity.
- Assess the possibility of permanent damage during drug washout and placebo studies in which participants are not taking medication.
- Carefully search the peer-reviewed medical literature.
- Explicitly discuss conflict of interest among IRB members.
- Hold explicit discussions with all sources who need to provide support for the IRB.
- Address financial support.
  - —How are participants to be compensated for costs of hospital care for adverse outcomes?
  - —How are participants to be compensated for irreversible severe adverse outcomes that occur during study participation and are related to the study?
- Help with questions and concerns related to the optimal protection of human participants of research.

Have All IRB Members Present for Voting

The strength of an IRB lies in having all of its members present for all discussion and voting, to ensure that all perspectives are heard before any vote is taken. The best IRB decision making is done with all members present, especially the nonscientists. The U.S. *Code of Federal Regulations* allows an IRB vote to be taken with a quorum present if it includes the nonscientist(s). However, in any close vote, the absence of one or more members can result in a decision that the majority of the IRB may not agree with. Take the case of an IRB with thirteen members. Twelve members are present at a meeting and vote six to six on whether or not to approve a particular research study; the thirteenth member would have voted nay. Because the IRB's operating procedures state that a majority vote is necessary, the chair postpones further discussion and voting until the next meeting. At that next meeting, three members (who would have voted nay) are absent. After a brief discussion, a new vote results in a six-to-four approval of the study. Thus, a research study is approved by a quorum vote, including the nonscientist(s), that would not have been approved if all members had been present. It was a legally valid vote, but was it an ethically valid one?

One way to improve the quality of voting by the majority of IRB members present at a session is to allow proxy voting, so that an absent member can still vote. Alternatively, all members might be present for discussion sessions but able to vote by proxy. But, because key new discussion may occur at any time along a study protocol's and informed consent form's evolution, the best approach is to have all members face to face at both discussion sessions and votes.

Have IRB Members Develop Areas of Expertise

In its efforts to protect study participants, the IRB must perform many tasks. One way for the IRB to manage these tasks is for individual members to develop particular areas of expertise.

As an IRB member develops expertise in an area, he or she may want to write a paper to add to the IRB literature, or he or she may do research on the issues that pose continuing conceptual challenges to IRBs. Such expertise has the potential of helping to clarify the issues that IRBs routinely face, like cognitive capacity, the uses and limits of placebo studies, the privacy of study participants and how best to maintain that privacy, especially in the case of genetic studies.

Careful Search of the Peer-Reviewed Medical Literature

All IRBs need help with searching the peer-reviewed scientific and medical literature to make certain that the risks section of a scientific protocol and informed consent form is complete and accurate. Some IRBs may be able to have their respective institutions' librarians help with searches. But the IRB member who is concerned with a particular topic will be able to specify the nuances of a point that should be researched. A member who has expertise in searching the peer-reviewed literature can play a vital role in enabling the IRB to know when principal investigators are doing well in stating risks and in helping the IRB find risks that have not been stated or were not stated clearly.

Types of Research an Institution Is Capable of Conducting Ethically and Safely

In order to get an understanding of what types of research a medical institution is capable of doing, one needs to understand (1) the clinical and research expertise of the principal investigators and co-investigators who will conduct the research, (2) the clinical expertise needed within a research study, especially one across disciplines, (3) the personhours available to do the study, and (4) whether it is safe to conduct the study in the particular context the study is being designed for. For example, if an IRB receives two new scientific protocols for emergent patient care, does the emergency department have enough experienced staff in place to deliver patient care and contribute to two research studies? What if one of the studies involves patients in acute exacerbation of mental illness, some of whom are arriving at the emergency room unaccompanied by a family member? An IRB member who knows the current wait time and staffing situation in the emergency department can help the IRB understand whether it is reasonable to add any research study to the current workload of that department. The IRB will need to decide whether such studies can be carried out safely in their institution with the optimal protection of the human subjects.

Cognitive Capacity of Potential Study Participants

In the area of study subjects' decision-making capacity, particularly subjects with the fluctuating cognitive states of delirium or acute mania or the progressive cognitive decline of Alzheimer disease, many questions have not been adequately addressed by researchers, clinicians, regulators, or ethicists. For example, a study in which individuals with fluctuating or declining cognitive capacity would be randomly assigned to receive either a new drug or placebo would generate many questions that the IRB member must consider and pose at a board meeting: How will the principal investigator justify to the IRB that an individual with fluctuating or declining cognitive capacity is giving voluntary consent? What criteria will the IRB use for its own decision making regarding the protection of study participants with fluctuating or declining cognitive capacity? What will be the roles of advance research directives and proxy decision making? What are the arguments for and against advance research directives and proxy decision making?

IRBs need to acquire expertise in measuring decisional capacity and in the use of the latest tools for assessing decision-making ability.

Scientific Issues versus Ethical Issues

The IRB evaluates each scientific protocol and informed consent form for both science and ethics. The vast array of scientific hypotheses involving human participants demands expertise in many distinct scientific and medical areas. Matters of opinion, debate, and dispute can be found in both the scientific and the ethical aspects of any study. If an IRB has five principal investigators who are studying the same or similar areas of science, there may be disagreement over whether the scientific design of the study is the best to answer the study question. There will always be disputes over whether particular study designs are optimal for the study of certain diseases and their treatments. An example of such disputes is discussions of the need to compare new study drugs not only to placebo but also to the best alternatives on the market. If a new study drug is compared to placebo only, then when it is marketed, clinicians have no information on how that drug compares to the best drugs already on the market.

Sometimes an IRB member focuses all the IRB's attention on the scientific question being raised to the exclusion of the ethical issues. This member may be attempting to overwhelm the IRB with the purported importance of the scientific question and failing to recognize that the study lacks adequate protection of participants.

A debate focusing on only the importance of the scientific question without considering whether the study design and research team can optimally protect participants, especially a debate that is being manipulated, can leave an IRB exhausted. An exhausted IRB may be inattentive and insensitive. Any IRB member who notices that a discussion has been unbalanced should call attention to what has not been discussed. Study participants deserve a penetrating review and discussion by the full IRB of how the principal investigator is proposing to protect them.

Although all IRB members should be observant of trends in the board's debate, the chair bears the primary responsibility for noticing and calling attention to unbalanced discussion. The chair must make certain that the nonscientists and new IRB members are following both the scientific and the ethical aspects of all discussions. No matter what the scientific issue is, the IRB chair must continually refocus the full board's attention on the optimal protection of study participants.

When the IRB reaches a decision, it must be well considered enough to withstand the challenges of its own institution, principal investigators and study sponsors, federal regulators, and others. Each IRB decision must be defensible against the following charges:

- The IRB did not understand the facts.
- The IRB drew incorrect conclusions.
- The IRB misunderstood the study and its goal.
- The IRB misunderstood what research is possible and what research is not possible within the patient care delivery constraints of its institution.

IRB members must be ready to defend their decision making when they feel it was correct, and if they suspect it was wrong to admit the fact, reconstruct how the decision making went wrong, and learn from their mistakes.

Explicitly Discuss Conflict of Interest among IRB Members

Any appearance of conflict of interest among members during discussions must be pointed out. An IRB member can turn the board's attention away from key topic areas by filibustering and not allowing other members to participate in the discussion. The IRB chair must be on the lookout for any block voting, in which several IRB members band together and vote the same way on a particular topic, intentionally taking a position against that held by the other IRB members.

The IRB membership itself can be manipulated to influence the way the IRB is voting. (In my experience, this manipulation is usually to promote research at the expense of optimal protection of study participants.) The IRB member must consider whether this practice is tolerable and if so to what extent, and whether he or she wants the subject discussed.

Conflict of interest can arise among members of the IRB when a member is serving as principal investigator for a proposal before the IRB, even though the member will have recused him- or herself from deliberations and voting about that proposal. The problem becomes even more complex when the IRB member has a particular area of expertise that the IRB relies on in its decision making. A typical result of such a conflict of interest is that the IRB treats the member's scientific protocol and informed consent form with less stringency than it applies to similar studies by other principal investigators. Cases may occur when one IRB member's proposal is reviewed more rigorously than another IRB member's, or is perceived to have been. The IRB must be consistent in applying review criteria across all research studies and all principal investigators.

Recognizing Problematic Studies

All principal investigators assert the importance of their research agenda. However, the IRB will not always be able to discern what a principal investigator is intending in the assertion. For example, let us consider a research proposal for a new study drug for a particular medical condition or disease for which there are already many approved therapies. IRB members reviewing the proposal might ask themselves the following questions:

- Did the principal investigator read the scientific protocol and informed consent form that were written by the drug manufacturer or simply submit them, sight unseen, to the IRB?
- Does this new study drug potentially have more benefits and fewer risks than drugs already approved and marketed for the same condition? If not, why is the drug being studied?
- Is the principal investigator being offered any financial incentives from the drug manufacturer that are not obviously stated in the materials submitted to the IRB?

Be Open with All Parties Who Support the IRB

The chair of the IRB and all IRB members must never hesitate to discuss any issue with the appropriate institutional officials. When disagreements arise between the IRB and other offices of the institution, they need to be discussed openly, because they will not go away. Do not accept arguments from administrators who do not understand human subject protection regulations as well as the IRB does. The institution's lawyers may not be acquainted with the regulations. The IRB must never hesitate to call on experts for opinions and to share those opinions with the institution's administration and counsel, who may well be grateful for the education. The more knowledgeable all parties are, the better the IRB can do its work.

Keep Accurate Meeting Minutes

The decision-making processes of the IRB can rarely be captured completely in the IRB's meeting minutes. It is desirable, however, for minutes to convey the nuances of the discussions and the exchanges of points of view that produce the board's solutions to the basic questions of how best to protect the study participants and whether or not the scientific gain is worth the risk.

The purpose of keeping IRB meeting minutes is to document the sufficiency and propriety of the decisions. This documentation is crucial in providing a history of the IRB's decision making and the grounds and criteria on which its decisions are based.

The documentation of discussions and decision making covers the range of IRB work and includes

- questions that come to the IRB from study participants, principal investigators, and anyone else
- point-counterpoint arguments made during review and evaluation of scientific protocols and informed consent forms
- notes from continuing review of approved studies
- training that is conducted to keep the IRB up to date

The meeting minutes provide a record of how the IRB functions over time with different members and different chairs. A detailed set of meeting minutes may serve as the only protection an IRB has when accusations are made against it. Minutes kept without attention to detail may become problematic when issues and disputes arise that can be resolved only by knowing what transpired, what arguments were made for which positions, and who took which position.

Keep Track of IRB Decision Making

All IRBs should keep a record of their discussions and decision making. They also need to keep track of the criteria they use in making decisions regarding research study participants in general and vulnerable participants in particular. Review of meeting minutes over a one- to two-year period often serves to highlight inconsistencies in criteria and in how they are being applied. Inconsistencies often can arise when the full board is not present at a meeting or when the members with particular areas of expertise are not present at a meeting. With this information, the IRB can reexamine the criteria it is using in its voting and its commitment in formulating decisions that are held by its full board.

An exercise that an IRB can set for itself is to develop a set of criteria for use in evaluating scientific protocols and informed consent forms and then reread the past five years of its meeting minutes to see if it has been consistent in applying these criteria.

Teaching through Review of Scientific Protocols and Informed Consent Forms

The continuous process of interaction of the IRB with principal investigators and others regarding the protection of human participants, if done appropriately, serves a vital teaching role. This process, when considered from the standpoint of time and work management, can significantly reduce the future workload of the IRB staff. Some of the most burdensome delays in the review and evaluation of scientific protocols and informed consent forms result from submission of materials that are improperly prepared and imprecisely written and when principal investigators give vague answers to the IRB's questions. This often happens when a principal investigator receives a scientific protocol and informed consent form from a study sponsor and submits them for IRB review and evaluation without taking the time to read and understand them. When the IRB raises questions, the principal investigator may simply glance at the areas referred to in the questions and make changes or argue against changes. Training of principal investigators that eliminates such behaviors may markedly reduce the questions the IRB has to ask a principal investigator in order to understand his or her scientific protocol and informed consent form.

Clear and Precise Written Communication

In its communication with principal investigators, the IRB must be committed to clear and precise phrasing of the issues and attention to detail by both parties. Clarity is especially important in IRB communications to a principal investigator who is not doing his or her work properly in writing a scientific protocol and informed consent form. If the IRB is imprecise in its questions to the principal investigator, this encourages the principal investigator to be imprecise in reply, which leads to frustration for both parties. Attention to phrasing and detail by the IRB in its written communications will model what the IRB expects from the principal investigator.

Requiring the principal investigator to clearly and fully explain the study to the IRB will help him or her explain it to prospective participants in the study.

Clear communication with all parties will also help the IRB develop a capacity to explain and defend its decisions.

Teaching Principal Investigators

Even though a medical institution's research service conducts mandatory training of principal investigators in the protection of human subjects, it is often difficult for the principal investigators to translate theory into practice until he or she receives detailed commentary and criticism of the scientific protocol and informed consent form from the IRB. The interactions between the IRB and the principal investigator also help IRB members to appreciate how little training many principal investigators receive.

The Problem of Teaching Study Sponsors and Product Manufacturers

Ideally, the training given principal investigators will have an educational effect on the study sponsors and product manufacturers when they write the initial scientific protocol and informed consent form. However, a study sponsor or product manufacturer can take the same proposal rejected by one institution with one IRB and submit it to another institution with another IRB in an attempt to gain approval without any modifications. Education of study sponsors and product manufacturers in how to optimally protect human participants is much more difficult than teaching principal investigators.

Study sponsors and principal investigators learn which IRBs are "easy" and which are "tough." The "tough" IRBs demand well-designed studies that optimally protect study participants. They look for research teams that are well trained in their duties to protect participants, protocols that ask appropriate study questions that may result in new and important generalizable knowledge, and sponsors that spare no cost in informing participants of the study risks and of the compensations that are available if participants sustain adverse outcomes.

A study sponsor, product manufacturer, or principal investigator that does not want to follow the IRB's recommendations may argue with the IRB. For example, they may assert that a group of study participants should not be considered vulnerable even though such patients are defined as vulnerable by the *Code of Federal Regulations.* Or they may argue that individuals may be offended if asked about their decision-making capacity or stigmatized by the use of proxy decision making on their behalf. The IRB must always be ready to defend its position. IRBs should take each such instance as an opportunity to teach study sponsors, product manufacturers, and principal investigators about the range of conflicts of interest in clinical research and the range of safeguards available to provide protection for participants.

14 Decision-Making Capacity and Accountability in Research

In the IRB's review of a scientific protocol and informed consent form, one of the key areas of scrutiny is how the principal investigator will assess a potential enrollee's capacity to make decisions. Decision-making capacity in research is particularly troublesome because of the amount of information that a study participant needs to understand and the breadth of that information, and because the study itself can have an impact on the person's decision-making capacity.

The most able decision maker can be challenged by the amount of information that needs to be understood to decide whether or not to enroll in a research study. It includes, among much else, data about

- foreseeable risk,
- unforeseeable risk,
- compensation for injury from adverse outcomes,
- compensation for disability from adverse outcomes,
- determining whether an injury will be considered related to the study, and
- legal rights of participants.

If the potential enrollee has existing medical or mental health conditions, these may affect the person's decision-making capacity. Among the conditions that can have such impact are

- sepsis (e.g., systemic infection),
- metabolic diseases (e.g., high blood sugar),
- moderate to severe depression, and
- acute severe mania.

Further research is needed on the effect that medical conditions and mood disorders can have on an individual's decision-making capacity.

Competence to Consent to Clinical Treatment and Research

Cognitive capacity is the ability to "understand, appreciate, reason."[1] Becky Cox White provides a basic definition of *task-oriented competence:* "A person is competent to do a task if the person knows what actions are required to complete the task and has the abilities needed to perform the task." White recognizes that it is difficult to separate "knowledge" from "capacity," because "knowledge partly depends on intellectual capacities" and these capacities "can improve with an increase in knowledge." For White, "capacities and knowledge are both necessary for competence, because competent performance requires the person to know what actions are required and how they should be performed and be able to perform them."[2]

White proposes an approach to her concept of "competency to consent" in clinical care: "Ultimately, competence must be defined by specifying its context, standard, appropriate capacities, and the role of consequences. Until these analyses are complete, neither clinicians nor bioethicists can know when decision-making authority is being properly assigned, which patients can accurately relate their own value structures to health care, and which moral principles govern individual therapeutic choices. To date none of these has been firmly established."[3]

A 1982 presidential commission held that "any determination of the capacity to decide on [or consent to or reject] a course of treatment must relate to the individual abilities of a patient, the requirements of the task at hand, and the consequences likely to flow from the decision."[4] For this commission, "decision making capacity requires, to greater or lesser degree: (1) possession of a set of values and goals; (2) the ability to communicate and to understand information; and (3) the ability to reason and to deliberate about one's choices."

Bioethicist Carl Elliott asks whether a person is to be judged incompetent by virtue of making a poor decision, making an irrational decision, or making a decision in an unsystematic, illogical, or erratic way. Elliott brings to the discussion of capacity the notion of accountability and further argues that "if a person is making a decision that will affect his or her life in momentous ways, we will naturally be concerned that he or she makes a sound decision. But because we recognize that a person generally has the right to make even unsound decisions, a judgment about competence ensures that whatever decision a person makes, it is truly his or her decision: a decision for which he or she can finally be held accountable."[5]

Cognitively Impaired Individuals

Some individuals have psychiatric diseases that are defined in part by cognitive compromise. For example, schizophrenia is defined in part by the notion that the patient has impaired judgment. Persons experiencing certain medical conditions (for example, in the throes of an acute stroke affecting cognition, with acute severe symptoms of shortness of breath or chest discomfort related to an ongoing myocardial infarction, and in metabolic states involving high blood sugar that needs to be treated) should be considered too compromised for the deliberation regarding participation in research.

Thomas Grisso and Paul Appelbaum have studied the abilities of patients to consent to psychiatric and medical treatments.[6] Appelbaum, a psychiatrist, noted that many research investigators feel justifiable concern over the decision-making competency of a wide variety of "non-mentally ill" patients, including geriatric patients, adolescents, and seriously medically ill patients, as well as those suffering from psychiatric disorders, "which by definition affect mentation." Appelbaum pointed out that "a recent study of decisional capacities in the inpatient treatment context demonstrated substantial impairments in approximately 52 percent of schizophrenic subjects and 24 percent of depressed subjects, but only 12 percent of seriously medically ill subjects."[7] The phrase "seriously medically ill subjects" refers to participants who were in inpatient treatment due to exacerbation of a medical condition, such as heart disease, in contrast to participants who were in inpatient treatment due to a mental health condition, such as schizophrenia or depression. For Appelbaum, "these data call into question . . . the ability of these subjects to protect their own interests in making decisions about entering research projects."

In all cases of individuals in a state of "cognitive compromise," a decision must be made whether they should be allowed to participate in research. This issue lies at the heart of an ethical dilemma: If scientific research is not done on persons in fixed and advancing severe mental health states (such as severe forms of schizophrenia), then where will the scientific advances in these disease states come from? Yet, it can be argued that research should not be conducted on persons in severe mental health states.

Richard J. Bonnie argues that current controversies over research with cognitively impaired individuals should be seen, in historical context, as a reminder of unfinished business in the regulation of research. For Bonnie, this unfinished business includes the need to properly balance participant "protection and methodological considerations in study design, the necessary links between administration of the research protocol and the human subject's ongoing clinical care, and means of ensuring that informed consent to research participation is provided by subjects or by their authorized surrogates."[8]

Severely Depressed Individuals

In the case of severely depressed individuals, Elliott notes that "most accounts of competence focus on intellectual capacity and abilities to reason, and depression is primarily a disorder of mood. According to conventional thinking, depression is primarily about despair, guilt, and a loss of motivation, while competence is about the ability to reason, to deliberate, to compare, and to evaluate." However, Elliott points out that depression "can impair a person's ability to evaluate risks and benefits. To put the matter simply, if a person is depressed, he or she may be *aware* that a protocol carries risks, but simply not *care* about those risks." If this perspective is correct, "IRBs need to take special precautions in allowing researchers to enroll depressed patients in research protocols."[9]

Assessment of Competence

The patient's family and significant others should be included in the decision process regarding whether a person whose decision making is compromised should consider enrolling in a research study. The family and significant others may be able to offer information that the principal investigator and research team do not have access to or that has not been adequately explained in the patient's medical records.

IRBs should consider independent review of the patient's situation and the patient's understanding of the research study's purpose, its nature (e.g., randomized versus nonrandomized, placebo versus nonplacebo), its risks, and compensation if injury occurs during the study. The IRB must decide what constitutes independent review and which experts are to be considered independent enough from the researchers and the medical institution to avoid conflict of interest on the part of the independent reviewer.

There may be a key role for questionnaires in checking persons' understanding of the terms and concepts used in the informed consent form and informed consent sessions. There may also be a role for a questionnaire in checking potential participants' understanding of their legal and ethical rights related to the study, especially the right to withdraw from participation at any time within the bounds of safety.

Understanding

Regarding informed consent in clinical care, Ruth Faden and Thomas Beauchamp describe a notion of "full understanding." For them, an individual has a full understanding if he or she apprehends "all the relevant propositions or statements that correctly describe (1) the nature of the action, and (2) the foreseeable consequences and possible outcomes that might follow as a result of performing and not performing the action."[10] Faden and Beauchamp do not require full understanding in the area of clinical care, but the informational requirements of informed consent in clinical research are much higher; an individual is agreeing to participate in a project from which he or she may or may not benefit, may be harmed, and may not be compensated for costs of hospitalization, long-term care, or permanent disability that may result from the harm.

The notion of understanding is one of the most problematic in regard to information used in securing informed consent for treatment or study participation. Often the principal investigator and study sponsor who want to recruit individuals into a study minimize the understanding required. Even in questionnaire studies, the issue of understanding is often complicated by researchers' desires not to inform the individual about some aspect of the study in which he or she is being asked to participate. Sometimes these arguments come from social scientists who believe that if the person understands too much about the scientific hypothesis to begin with he or she could "game" his or her responses toward or against the hypothesis. The IRB must thoroughly consider why a principal investigator is requesting to underinform an individual about a study. The IRB's work is to help clarify the research, not to allow individuals to be misled or to misunderstand the research. The IRB must urge the principal investigator not to misrepresent the study hypothesis and, if necessary, to develop a new hypothesis and new research methods that do not involve any misrepresentation or deception.

Grisso and Appelbaum identify four standards used in the assessment of decision-making capacity in patients in clinical care in the United States:[11]

- expression of a choice by a patient,
- understanding of information relevant to a choice,
- appreciation of the significance of the choice for one's own situation regarding illness and possible treatments, and
- ability to manipulate the disclosed information rationally (reasoning about the information) in a manner that allows one to make comparisons and weigh options.

The notion of appreciation—that is, whether or not an individual understands that the information in an informed consent form applies to him or her —is equally problematic. How is one to assess appreciation? IRBs are continually challenged in the attempt to conduct research in areas of medicine in which individuals have conditions, such as severe depression and acute severe mania, that may affect cognition; but research about these conditions is needed, because of the limited effectiveness and safety of currently approved therapies.

Therapeutic Misconception

Prior discussion and education are at the heart of optimally informing individuals about participation in research. But even if a research study has been carefully developed and the study hypotheses have been clearly and precisely presented in the scientific protocol and informed consent form, individuals considering participation may not understand what the research is about and the difference between research and clinical care. This problem of "therapeutic misconception" has implications for decision-making capacity: does an individual have the decision-making capacity to participate in research if he or she fails to understand that the study is designed not to *provide* therapy but rather to assess a new therapy to see if knowledge can be acquired that will be useful in the treatment of future populations?

The IRB has a role in helping principal investigators clarify the distinction between research and clinical care, and the distinction should be made in an introductory statement in all informed consent forms. Any individual considering research participation should be given an informed consent form that explicitly defines the concepts of research and clinical care before explaining the specifics of the particular study.

The medical institution should foster individuals' understanding of the separation between research and clinical care. With IRB assistance, the institution can develop educational programs and produce posters, fliers, and handouts explaining the difference between participating in research and receiving clinical care at that institution. The posters can be displayed where announcements promote opportunities to participate in research. Each point of contact in the recruitment of research participants is an opportunity to emphasize the distinction between research and clinical care.

The problem is that the ambiguities that exist in an individual's mind over what is research and what is clinical care probably cause more participants to enroll than would sign up if they fully understood these terms. The IRB protects study participants when it helps to ensure that the individuals being recruited understand what it means to participate in a research study. These concepts will continue to evolve, and more research is needed on the best ways to educate individuals regarding the distinction between research and clinical care.

Advance Discussion about Participation in Research

All potential study recruits should discuss the decision with family members and significant others, and they may consider appointing a surrogate decision maker to be consulted regarding decisions about participation in future research projects if the patient is in an impaired cognitive state before the study or may become so during the study. The IRB must make certain that whomever the patient selects as a surrogate decision maker follows both state and federal regulations for someone entrusted with that authority. The advance research directive and the surrogate decision maker must be carefully considered by the participant, and, probably, the choices should be renewed or reaffirmed over time, to make certain that the participant has not changed his or her mind with regard to the advance directive or the surrogate decision maker.

In deliberations about advance research directives and surrogate decision makers, the advance directive frameworks that have been established regarding drug treatments for patients with schizophrenia will prove useful. For example, a patient with schizophrenia may have been treated with various medications and comes to prefer certain treatments and to dislike others according to the side effects and adverse outcomes experienced. This patient, then, on the basis of experience with various treatments, constructs an advance directive specifying his or her preferences for therapy. The directive can be used when the person is in a compromised state. In the research circumstance, how can a person know what he or she would want if in a state of cognitive compromise? What would the person's answer to this question be if he or she were in mild to moderate pain as opposed to being pain free? These are the questions we face when we consider the viability of advance directives regarding participation in future research studies.

It may be difficult at any one time to be sure that an individual with waxing and waning awareness has adequate capacity to consent to participate in research. When an individual is not cognitively impaired, he or she may be able to discuss and deliberate about the construction of an advance research directive, which details precise situations in which the person allows the consideration of participation in research. The best time to approach the cognitively impaired or compromised person is either when his or her cognitive state is unimpaired or when his or her state of mind is, to reasonable observers, in a state of competency or decisional capacity. (The term *competency* tends to be used in legal contexts and *decisional capacity* in medical ones.)

If there is a question of whether an advance research directive exists and whether it should be acted on, one should ask next of kin or significant others their opinion. Next of kin or significant others may be able to confirm that an advance research directive exists and that it was constructed and signed without coercion when the person was not impaired and that it reflected the person's wishes. *Even if an advance directive is in place, with agreement by a surrogate, proxy, or legal representative, if an individual with waxing and waning capacity declares that he or she does not want to participate in research, this statement must be taken as valid, and it countermands anything in the advance research directive or said by the surrogate, proxy, or legal representative.*

Research about Mental Illnesses

IRB members will be asked to review research related not only to medical conditions but also to mental health. Cognitively compromised individuals often will be among the participants in such research. When is the individual with a severe medical condition or a mental illness that influences his or her cognition and mood to be considered as a potential participant in a research study? Should the institution get involved in such studies at all? A few of the questions IRB members should ask are listed in Box 14.1.

Box 14.1. When the Study Is Examining a Mental Illness

- Is the individual being approached for participation in a condition that may require urgent or emergent care? If so, will the IRB allow recruitment if there is no existing research advance directive and no requirement for a surrogate decision maker?
- Under what circumstances, if at all, will the IRB allow research study of a mental health state (e.g., depression or mania) in its institution?
- If the mental health state being studied is depression, how will the IRB optimally assess whether an individual is agreeing to participate not because he or she has evaluated the study and made a voluntary choice but simply because he or she does not care?
- If the mental health state being studied is acute severe mania, how will the IRB optimally assess whether an individual not only understands the facts related to participation in the study but also appreciates the fact that he or she is in the state being studied and that he or she will be participating in a study of that state?
- What stand will an IRB take regarding a study comparing a new drug to a placebo when there is incomplete information as to whether permanent damage can occur while the patient has washed out the drug he or she was receiving and is taking the placebo?

Consideration must be given to additional methods of protecting the especially vulnerable participants who would be enrolled in these studies. Such methods include advance discussions with the individual concerning participation in research (as in a research advance directive) as well as the involvement of surrogate decision makers who have been appropriately selected on the person's behalf.

Willingness to Take and to Accept Risk

One of the relatively unexplored areas of risk in IRB work and in research on human subjects is cognitive psychological research into willingness to take and to accept risk. While cognitive psychologists have begun to examine the differences between individuals in their willingness to take risks, little work has been done into an individual's willingness to accept a risk that befalls him or her, as during a research study. Although an individual may report being willing to take risk when risk is viewed theoretically, he or she may not be willing to accept an adverse outcome when the risk materializes in his or her particular case. The basic question is, Does the individual have the decision-making capacity to determine whether or not he or she is willing to participate in a research study that may (or may not) lead to generalizable knowledge for the health and care of future patients and may cause the individual harm (sometimes severe) without any chance of benefit? To begin to answer this question, the IRB member must ask these questions:

- Is it clear that the individual being approached has the decision-making capacity to participate in research?
- If an individual has a research advance directive in place,
  —does it appear that the individual adequately considered the content of that research advance directive explicitly enough to understand what he or she is getting into in a high-risk study?
  —is it known where this research advance directive came from?
  —did the IRB ever see it? If so, did the IRB approve it? If approved, what criteria did the IRB use to evaluate its contents and when it was to be used or applied?
  —how will the IRB evaluate the research advance directive and whether the individual truly understood its content before signing it?
  —how would the IRB know if the individual changed his or her mind after signing the directive but did not know how to rescind it?
- If an individual has designated a surrogate, is it clear that the surrogate wants to take on the responsibilities of decision making, including in genetic studies that can have an impact on the participant's past, present, and future relatives.

Genetic studies need special attention, because what is being consented to is so much broader. For example, there is a risk that stored genetic materials may be mislabeled, and the mislabeling may at some future time misidentify a participant or a participant's family member as having a genetic disease or a genetic marker for a disease. Uniquely identified specimens pose risks that must be discussed fully and clearly with participants who allow their tissues to be uniquely identified and stored for future genetic studies.

Where there is risk there is potential for liability. Perhaps the most difficult part of the informed consent form for any study participant to understand is the liability section. A prospective enrollee needs to understand who will pay for hospital costs and costs of compensation if a severe adverse outcome befalls him or her. The liability section may state, "The study sponsor does not offer compensation," but it may not explain that, in a legal battle to determine liability after a severe adverse outcome has happened to a participant, the study sponsor may argue that the adverse outcome was not related to the study. The informed consent form may state, "The study sponsor will pay hospital costs for study-related injury occurring to participants in the sponsor's clinical trial," but it may not describe the problems participants can encounter in attempting to determine whether a severe injury that occurred during the course of a study was related to the study.

Rarely if ever does an informed consent form discuss the ambiguity of the ways in which causality is determined for a severe injury that occurs during study participation and the fact that the participant may have to take the case to court for adjudication. A participant's willingness to take and accept the risk of participating in a research study may be assumed when he or she signs the informed consent form. However, the details of what happens if that participant sustains a severe injury and the battles that can sometimes ensue in assigning causation and determining whether the injury was study related are rarely if ever discussed in the clear and precise terms that should appear in informed consent in clinical research.

Competence and Accountability of the Participant

Competence was first taken on as an issue to be clarified in law and medicine within the context of clinical care. In the case of an emergency and without any advance medical directive by the patient, a physician would likely initiate on the patient's behalf an intervention to preserve the patient's life. In the absence of an emergency, a physician could not proceed with an intervention without the patient's consent to be treated. There was a need to develop and adopt a set of procedures to evaluate the patient's capacity to accept or reject an intervention recommended by his or her physician and a need for a patient to have the capacity to decide whether he or she wanted to accept or reject a proposed intervention. Capacity was taken to mean an individual's abilities to understand the situation he or she is in with respect to health at a particular time and whether or not he or she wants to be treated.

The capacity needed to make decisions regarding participation in research is a much deeper requirement than that regarding clinical care. That is why the Belmont Report called for a new standard, beyond the professional standard and the reasonable person standard (see Chapter 10). The report emphasized that this new standard was especially needed in research on patients already receiving or in need of treatment, particularly for a severe disease affecting his or her life. Such a patient is coming to the medical center, hospital, or clinic not to become a participant in a research study but to receive clinical care. The recruitment of that patient into an ongoing research study slows the delivery of clinical care. Time is taken away from clinical care to decide whether or not the patient can be a part of the study (that is, meets the inclusion criteria) and then the patient is approached for informed consent. Because research in the clinical setting takes the patient away from clinical care (or at least away from an aspect of clinical care), the decision to participate in research is difficult, even more so if the patient's condition needs emergent care. The status of an individual who is seeking urgent care at a medical center for the management and evaluation of a worsening mental state may be significantly compromised in his or her ability to make any decision regarding clinical care, let alone any on-the-spot decision regarding participation in a research study.

The decision-making capacity of study participants, especially vulnerable participants, needs research. It should be approached in a multidisciplinary fashion involving clinicians, communication specialists, decision analysts, ethicists, researchers (especially those in the cognitive and behavioral sciences), and translators, among others. Of particular importance is the study of individual decision-making capacity in the presence of mental health conditions and other medical conditions that can affect such capacity.

Much of the focus of the assessment of decision-making capacity has been on what is in an informed consent form, testing for understanding of facts and concepts and reasoning, yet bioethicist Carl Elliott called attention to additional key factors that are only beginning to be assessed systematically: emotional factors. He pointed out that IRBs must draw principal investigators' attention to how a person's affective and motivational state affect decision making. He described "the depressed patient who is capable of understanding all the facts about his or her illness and the research protocol in which he or she is enrolling and the broader implication of the protocol in his or her life, but who, as a result of his or her illness, is not *motivated* to take those risks into account in the same way as the rest of us." As Elliott says, "these patients might realize that a protocol involves risks, but simply not *care* about the risks. Some patients, as a result of their depression, may even *want* to take risks."[12]

Elliott introduces a further aspect of competence or decision-making capacity—accountability. He uses the example of a patient with mania: "If a person were to behave badly while mentally ill—say, in a full-blown manic episode—we would likely think it unfair to hold him or her fully responsible for what he or she has done. His or her behavior was uncharacteristic; he or she would never have acted this way unless he was manic; his or her mania is temporary and reversible with lithium treatment. His or her actions in the manic state were not truly *his or hers*."[13]

Elliott extends the example to the patient with depression. "Similarly for the depressed patient being asked to consent to research: his or her mental state is such that his or her behavior and choices do not seem to be truly his or hers. If something untoward were to happen to him or her during the research, for instance, we could not in good conscience say that he or she bears the full responsibility for undergoing that task." For Elliott, "the authenticity argument supposes a person with a stable character and entrenched dispositions, who, while depressed, makes decisions so uncharacteristic that we feel bound to question whether those decisions were truly his or hers."[14]

If a patient in a state of acute severe mania or acute severe depression is coming to a hospital for care for the first time by himself or herself, how could the patient be considered to have the decision-making capacity to understand and appreciate the facts presented and the applicability of those facts to his or her own circumstance and current state of mental health? Is this individual to be considered accountable for an informed consent to participate in a research study? The continual search for what it means for a study participant to be accountable for a choice and decision making in regard to research participation is one of the IRB's main tasks in the protection of participants.

We now see the importance of two issues beyond understanding the information presented to a study recruit: appreciation (understanding not only the facts related to the research but also that one has the condition or disease that the research is about and understanding that the risks related to research participation apply to oneself) and accountability (understanding that one will be held accountable for the decision to participate in the study for as long as one is willing to participate).

Accountability of Parties Other than the Participant

In responsible research, there must be accountability at every level. The IRB holds the principal investigator and study sponsor accountable, but the concept of accountability applies also to the research team, the institution in which the study will take place, the research service that has allowed the study to take place, and the IRB that has approved the project and follows it over time.

The IRB is accountable for the quality of its own decision-making ability. The more conflicts of interest on an IRB and the more decisions that are made by less than the full board, the fewer optimal decisions are made. Decisions should be made by the full board and should involve relevant experts on particular questions. All members should have the opportunity to contribute to a decision for which the IRB will be held responsible. The only reason for the full board's not participating is if a member must recuse him- or herself because of a conflict of interest. Given the responsibilities of protecting human participants and recognizing conflicts of interest on the IRB, there can be no short cuts in IRB decision making.

IRBs are held accountable for ensuring that the principal investigator and research team optimally protect study participants. If the IRB is told about or identifies any problem in protecting participants during its reviews, observations, or discussions, the IRB must intervene, determining how to protect the participants, suspending the study until corrective actions have been implemented, or terminating the study.

Elliott's issues for principal investigators to attend to also apply to the IRB. The IRB must ensure that the individual being recruited into a research study is also assessed regarding how his or her mood is affecting decision making, "whether a person's mood has dramatically changed recently," whether the individual has a concern for his or her own well-being, whether the individual examines risks and benefits, and how carefully the individual examines risks and benefits. The IRB should foster research into these issues and encourage researchers to study the impact of emotions and mood on decisions to participate in research. Elliott calls for a reexamination of researchers' arguments that "exposing competent adults to these sorts of risks [for example, the worsening of a patients' depression as a result of being randomly assigned to the placebo group in a study] is justified at least in part by the institution of informed consent, with the corresponding presumption that potential participants understand the risks of the protocol, consent to them, and can be judged accountable for undertaking them. But if severely depressed patients are not competent, exposing them to the risk of having their illness worsen is much more difficult to justify."[15] IRBs must protect participants and must focus the principal investigator's attention on the gaps between understanding the concept of decision-making capacity and understanding the impact of emotion and mood on decision making. More research is needed to clarify both decision-making capacity and the impact of emotion and mood on decision making.

Accountability in Reporting of Adverse Outcomes

The form in which the principal investigator, study coordinator, and research staff communicate information to the IRB has an effect on the IRB's decision-making capacity. In particular, the reporting of adverse outcomes, their severity, and the chance that they were sustained through participation in the research study must provide clear and precise information in enough detail for the IRB to act appropriately to deal with the consequences and prevent recurrence of the adverse outcome if possible. Often, reports of adverse outcomes are short and too simplified for the IRB to determine whether to suspend or terminate a study until more careful analyses are completed. One can imagine an adverse outcome report concerning a bout of cholecystitis that resulted in the death of a participant: "Adverse outcome: cholecystitis. Chance due to research drug (no chance, possible, probable, definite): possible." One missing element is any description of the event: "Adverse outcome: bleeding." Yet, the descriptor *bleeding* does little to help the IRB in its tasks and decision making related to the adverse outcome.

In cases like this hypothetical one, the IRB would need to know the answers to these questions:

- Why did the bleeding occur?
- Was the participant already at risk for bleeding and therefore inappropriately included in the study? (If so, the interview procedures or inclusion and exclusion criteria may need to be changed to better identify individuals with a risk of bleeding or tendency toward bleeding.)
- What happened to the participant as a result of the bleeding? (All of the consequences of the bleeding must be clearly specified.)
  —Did the participant have to be hospitalized in an intensive care unit?
  —Did the bleeding require surgical intervention?
  —How many days of hospitalization were needed?
  —Is the study sponsor paying for the hospitalization?
  —Did the bleeding cause any temporary or permanent damage (for example, stroke due to hypotension from blood loss for a lengthy time)?

- Was the bleeding due to an error or an oversight by the research team? If so:
  - —How has the research team been trained to correct this oversight?
  - —What plan of operation has been put into place to ensure that such an error or oversight never occurs again?
- Could the IRB have anticipated this adverse outcome and its possible sequelae in its initial review of the scientific protocol and informed consent form?
- Could the IRB have done a better job of reviewing the mechanism of action of the study drug and its possible relationship to bleeding?
  - —What inclusion/exclusion criteria were being stated in the research protocol?
  - —If the drug was previously studied or if there are "like drugs" on the market which have bleeding as an adverse outcome, what risks were reported in the peer-reviewed medical literature?
  - —Who was in charge of identifying laboratory abnormalities?
  - —Was the time period between the laboratory tests and the review of the laboratory test results prompt enough to have picked up an abnormality before the bleeding?

The basic approach to providing information about an adverse outcome is described in Box 14.2. The IRB member can expand this list as needed for the particular situation.

Box 14.2. Evaluating an Adverse Outcome

- What was the nature of the adverse outcome?
- Did the adverse outcome result in death, stroke, or other severe harm to the participant?
- Describe the precise event the participant experienced.
- What is the numerical chance that the adverse outcome was caused by, associated with, or related to participation in the study?
- How may evaluators (estimators) agree that the adverse outcome was related to the study?
- How many evaluators (estimators) disagree that the adverse outcome was related to the study?
- List the disagreements among the evaluators (estimators).

- What are the qualifications and relationships of those providing this description and estimate?
- Describe any conflict of interest of those making judgments as to whether or not the adverse outcome was related to the study.

The IRB should develop an outline for detailed reporting that will provide adequate information about adverse outcomes, their severity, why they occurred, and what will be done in the future to minimize similar occurrences and to allow earlier detection and notification and guidance of participants. The IRB's directions should include a caution about numerical estimates and measurements. Any use of numerical scales should be accompanied by verbal descriptors, so that all IRB members and any participant, regulator, administrator, or attorney who will read the report can easily grasp the meaning. (This is an area of ongoing investigation: how do you best present numbers to people who are not comfortable with numbers?)

Accountability of the Institution

Although the IRB is an institution's specialized monitor of research going on under its auspices, the institution's chief executive, the chief of staff, the administrators, the clinical service chiefs and hospital clinical staffs, and the research service personnel all must understand what is involved in protecting human research subjects. The IRB must keep meticulous records and must clearly communicate to the institution's administrators what it is doing to protect participants (especially if some people in the institution are more interested in conducting research than in protecting participants) and what the administration can do to facilitate the goal of protecting participants. The IRB must establish communication as the norm, sending meeting minutes to all relevant parties. An IRB can educate others in the institution about research and the best protection of participants. It should aim to ensure that no one in the institution can say that he or she does not know how to communicate with the IRB or learn its policies and recommendations.

When an important or generalized problem with the institution's studies has been identified and acted on, the IRB must make certain that all researchers working in the institution understand what the problem was and the best way to respond to such a problem should it arise with one of their studies. If there are problematic researchers or research staff members who fail to change their behaviors, then counseling, education, and training should be obtained or those persons not permitted to be involved further in research at the institution.

The IRB must be receptive to constructive feedback from all research and institutional personnel. All issues should be discussed fully and with clear focus on the institution's task of protecting participants. Principal investigators coming from another institution may not have received general or specific training in the protection of human participants, may have interacted with IRBs, research services, and institutions with looser standards than this one, and may have made mistakes that were not discovered at the previous institution.

Both the principal investigator and the medical institution benefit from monies coming in to finance research, both benefit from the prestige that comes from research discovery, and both must maintain the public trust to sustain research on human participants as a viable endeavor. These benefits can be sustained over time only if the institution can assure individuals that they will be optimally protected during participation in a research study.

Institutional accountability is of particular importance in the realm of innovative procedures and devices. Their use may not strictly be research, but the protection of patients must be the paramount concern of the institution and its staff. Let us take an example. Although interventional radiologists use FDA-approved devices, not all approved devices are designed to fit all patients. Emergencies arise when an interventional radiologist may need to modify an existing device or create a new device to fit a patient's unique anatomy. The radiologist must document the adaptation that was made. The question may then arise whether the modified or new device might be useful for other patients in the future. At this point, the medical institution, in consultation with regulatory agencies, has to set the ground rules for transitioning this emergent device into a formal research protocol. With the transition will come the procedures to ensure the optimal protection and informing of patients about the experimental and research nature of the device or procedure. Scientific protocols and informed consent forms will have to be developed under the oversight of the IRB, with the knowledge of any clinical service chief involved in the use of the device in the medical center, and with clear and early communication with the FDA regarding the adaptation. All principal investigators and clinical staff must be fully informed of how to report to their relevant clinical service, research service, chief of staff, and hospital administrators any new procedure developed during emergent or urgent care.

Avoidance of Influence over IRB Decisions

The IRB must always be aware that, intentionally or unintentionally, other parties can influence IRB decision making, for instance, away from the protection of study participants and toward securing research funds for the institution.

Such influence may come in many forms: the appointment to the IRB of an individual with an agenda, the selection of a speaker who will present to the IRB only one side of an argument, or the calling in of an expert on only one side of an issue. Even the formation of a subcommittee of the IRB, inappropriately recommended by an institution's administration, might be aimed at speeding up the IRB's decision making to generate more research projects. Any time issues are discussed without the full board being present, an opinion held by a subcommittee but not held by the full IRB could be imposed on the decision. The strength of an IRB lies in the diversity of its members and in hearing all sides of arguments, with the best representatives of all sides presenting and enough time given for careful deliberations by the full board.

The Decision-Making Capacity of Individual IRB Members

The IRB chair and members will always be a subject of analysis and judgment within the medical or research institution. They bear a lot of responsibility as the primary guarantors of the institution's protection of research participants. This role may place the IRB chair and members in conflict with the research service and others in the institution, because the IRB's goal is to protect participants, whereas the research service and medical institution may be more inclined to promote research. When problems arise, most research services and medical institutions, perhaps being defensive, will deflect attention to the IRB chair and members, charging them with not preemptively recognizing the problem.

The recommendations in this book are based on the assumption that the IRB already has a preponderance of members whose primary aim is the protection of human study participants. The decision-making of the IRB should be called into question any time the IRB becomes a force for the promotion of research or is repeatedly conflicted in this area. If this occurs, the best action is to remove from the IRB those members whose aim is the promotion of research at the expense of the protection of human participants. The institution may choose to form a separate committee charged with promoting research, but it must have no authority over the IRB.

Research is needed on the composition and functioning of IRBs and their focus on the protection of human participants and on the best ways to retain and develop that mission.

Summary

The IRB's Key Role

The IRB is the cornerstone of the enterprise of protecting participants in clinical research conducted by U.S. investigators both in and outside of the United States. Here is a review of the IRB's tasks and the environment in which it performs them.

Regulations and Guidance in the Primary Task

IRBs must follow the U.S. *Code of Federal Regulations,* yet these regulations are subject to interpretation. There are federal regulators who provide oversight of and guidance to IRBs. When an oversight group comes to a medical institution to evaluate the work of an IRB in a particular study that has been called into question, the group will usually review the entire functioning of the IRB and provide guidance in the areas in which the regulators believe that the IRB is not functioning satisfactorily.

Clearly, the best way to protect study participants—the primary task of any IRB—is for the IRB member to keep a continuous focus on protecting participants as he or she reviews all scientific protocols and informed consent forms, deliberates on all relevant issues, and discusses each point with the rest of the board. The IRB member must continuously ask the questions: What can go wrong? What harms can be done? How can the IRB best intervene now to help prevent harms? How can the IRB help ensure that the research team is best qualified and best constructed to carry out a study with minimal harm and, if any harm should occur, identify and act on it as early as possible to minimize the consequences for the participant?

Each IRB member must speak up when he or she does not understand something in a scientific protocol or informed consent form. Members should not ignore any issue that needs clarification. They should ask the principal investigator clear questions about the problems and not let the principal investigator fail to clarify them. A problem one IRB member has, for example, with a scientific protocol or an informed consent form is usually a problem most of the IRB members recognize but may be reluctant to speak up about. Once one IRB member has the courage to bring up the issue, there usually will be agreement that the issue should be addressed. The principal investigator is then asked to make the needed clarification or correction, to reformulate the scientific protocol or informed consent form if need be, and must accept rejection if he or she fails to offer the needed modification or revision.

When there are federal or state laws that bear on a study that an IRB is considering, the IRB must obtain opinions from counsel. The IRB must be deeply familiar with federal and state laws regarding research on human participants, especially in the area of research on uniquely identifiable specimens in genetic studies, where new laws are being made.

Increased attention is being paid to the entire medical and research institution within which an IRB functions. People at all levels of the institution must understand research and the protection of study participants. Because IRB members deal with these issues on a daily basis, they are well-equipped to help educate all relevant parties within and outside of the institution regarding the best ways to protect participants. For clinical research to succeed in the long term, the institution where it goes on must have an overall climate of interest in protecting participants.

Understanding the Science and Ethics of Research on Humans

IRB members examine both the scientific issues involved in the research study the IRB is reviewing and the ethical issues related to the protection of human participants. Searching the peer-reviewed scientific and medical literatures must become a routine part of an IRB member's fulfillment of the role. Unless the IRB has a clear method of determining when to search the literature, the task may be considered burdensome. Many IRBs will need to seek assistance from their institution in performing thorough searches of the literature (e.g., medical librarians, medical staff members trained to access the literature on research topics) so the most current knowledge is brought to bear on each scientific protocol considered. In addition, the IRB must seek out local, regional, and national experts who can help it understand the science involved in each research protocol.

The IRB's commitment to science and ethics in the review of research protocols and informed consent forms also requires a commitment to learning about statistics and statistical interpretation. IRB members need to be conversant with key scientific points about the number of participants needed in particular types of studies. They also need to become skilled at recognizing those studies that have other-than-scientific goals, for example, a product manufacturer seeking to establish a foothold in a hospital for a new drug or for a new use of an old drug.

Evaluating the Research Team

The IRB has many opportunities to take the measure of the principal investigator, the person responsible for presenting the scientific protocol and informed consent form to the IRB, defending them, and repairing them under the IRB's direction. The IRB must also look closely at the research team's qualifications to conduct the study. The board must say so if it feels that more researchers or more-qualified researchers should be added, making certain that the research team includes enough qualified personnel to ensure that study participants are protected. The protection of study subjects is most clearly needed when researchers are working on a study outside of their field of expertise. If a study crosses two services (for example, surgery and internal medicine, or psychiatry and internal medicine), there must be investigators or collaborators from both services, to ensure that each participant is being properly cared for during the research and to ensure optimal communication with the participant's primary physician.

The Protection of Participants in Multisite Studies

If a research project will be conducted at more than one institution, the IRB at each site needs to know how participants will be protected both inside and outside their own institution at each point in the research process. Participants will need ready access to the principal investigator and the research team and also to the IRB itself, to obtain answers to questions about how the research is being conducted.

An IRB should look at how other IRBs are treating the same scientific protocol and should share information and approaches with other IRBs.

Assessing Individuals' Understanding of Participation in Research

We are never certain how much an individual considering study participation understands about the scientific protocol and study-related risks, even if we do extensive paper-and-pencil testing (asking participants questions concerning what they understand and what they are confused about) during the informed consent session and at later intervals. There are two types of assessment: a more objective assessment (for example, by a questionnaire study) and a more subjective assessment (for example, by a one-on-one interview conducted by a highly trained professional).

One of the more difficult tasks in the assessment of an individual considering participation in a research study is to address the degree of understanding of an informed consent form. In research today, traditional concerns related to informed consent (for example, what it means for an individual to "understand" a scientific protocol or informed consent form) have broadened to include such notions as "caring whether or not they participate" and "whether they appreciate that they actually have the condition, disease, or disorder being studied."

Distinguishing between Treatment and Research

The participant in a research study must be informed that he or she is involved in a research study and not in clinical care. There are many reasons this distinction can be difficult to grasp, and over the course of time it can be lost again, for example, as a participant settles into a routine of visits to the medical facility and becomes familiar with the project staff. Reconsent is one way of combating any misconception that the study is actually treatment. Reconsent is also a way to remind the participant that he or she has a right to withdraw from the study, within the bounds of safety. Such reminders of the experimental nature of the study are necessary in part because, if the participant has an adverse reaction to a drug or device and needs to seek urgent or emergent care, he or she will have to notify the care facility that he or she is enrolled in a research study. The participant will need a method of contacting the responsible principal investigator(s) and staff to obtain answers to any questions regarding study participation and any adverse symptoms or signs. Other approaches to reminding participants of their continued enrollment in a study need to be studied.

Record Keeping

Each individual considering research participation and each study participant should have a written record of each encounter with the research team. Such a record would have at least two sections. The first section would contain the definition of research and how research contrasts with clinical care. Each definition should be illustrated with examples and problems that participants may face in deciding what is being done in their lives at the medical institution that relates to research and what relates to clinical care. The second section of the record would contain the scientific protocol and the informed consent form that the individual signed on entry into the study. There might be additional sections, such as a table of contents, a glossary, and an index. The participant would have this record to show the primary care or subspecialty physician caring for him or her outside of the study and for future research purposes. The participant can bring the record to each research session for note taking and would add to it as changes in the informed consent forms are made, reconsents occur, and summary reports come in.

The same level of meticulousness in record keeping must be maintained by all parties involved with a research study. Such records are especially important should any adverse outcomes occur, for there will be an immediate need to determine the cause of the harm to the participant, how it is to be handled, and how further similar occurrences are to be prevented. In the event of a law suit over compensation, accurate records will be necessary.

For all parties in research projects, records of what has transpired in a study enable learning from experience and expansion of skills.

Training and Accountability of IRB Members

There must be initial and continuous training of IRB members so that they can best fulfill their roles in an ever-expanding world of research hypotheses. At least initially, IRB members usually do closer critiques of studies outside their own area of research. Eventually, they are able to scrutinize their own and their colleagues' scientific protocols and informed consent forms with the same level of intensity. Researchers called to serve on an IRB will build on their experience and should become model researchers and advisors within their own department or service, better able to design their own research projects and those of their colleagues for the best protection of human subjects.

IRBs need to systematically analyze their performance of the tasks they undertake, assessing which procedures work and which do not. The training of IRB members and the recording and analysis of its deliberations will pay off in many ways. One is that the IRB itself may become a subject of a lawsuit, so it must be prepared to defend its decisions and how its decisions are made.

The IRB must be attentive to new approaches in areas like the assessment of decision-making capacity, continually updating their knowledge of these issues as new research is published. IRB members must continually focus on developing a better understanding of what they are doing in protecting participants, what criteria they are using to evaluate science and ethics, and whether they are applying sound criteria across all studies to ensure that they are treating all studies with the same degree of scrutiny and forethought as to their main purpose: the best protection of study participants at all times.

Goals

This book is intended as an overview of tasks that an IRB member could expect to encounter in attempting a thorough and effective review of a scientific protocol and informed consent form. Although the institutional review board is the first and last line of defense in the protection of study participants, the IRB member is not alone in the review process. The institution's research service is equally responsible for assuring that the IRB conducts thorough reviews. The research service's responsibility for thorough and systematic review includes reviewing the IRB's minutes, providing staff members to assist the IRB and helping the IRB in other ways that fall into its areas of expertise, including medical, surgical, and mental health care, law and ethics, communications, and searching peer-reviewed literature.

IRB review is challenging. For many of the issues faced in the process there are at least two sides: what the reasonable study participant would want to know, and what the principal investigator wants to disclose. The IRB often needs to do its own research on issues presented in the scientific protocol and informed consent form, issues as varied as baseline measurements, appropriateness of inclusion and exclusion criteria, riskiness of the intervention, appropriateness of language, and personnel on the research team. With experience, IRB members will go further in the analysis of specific topics than I have suggested in this book. By examining each scientific protocol and informed consent form in such detail, the IRB helps enable the individual considering participation in research to understand as much as possible about the particular research endeavor which he or she will be assisting.

Protection of patients and human research subjects has taken some great strides from its beginnings. The ethical principles of nonmaleficence, beneficence, autonomy, and quality of life are much better understood as basic concepts in clinical care.[1] Yet, many issues remain that are of both theoretical and practical importance. How are the scientific and ethical principles involving the protection of participants to be further developed and refined, not only from theoretical perspectives, but also in practical aspects? How are participants, from recruitment to the completion of research participation, to be optimally protected within the context of clear and honest communication about research?

The optimal protection of participants in clinical research has a long way to go. Some questions are ripe for discussion: How does one optimize the protection of participants at an institution geared towards promoting research? Is the only criterion that distinguishes clinical research from clinical care the acquisition of general knowledge? How much difference should exist in the decision making of IRBs at different institutions? Given that the guidelines in *The Code of Federal Regulations* are subject to interpretation, who is making these interpretations at each institution, and are there significant differences in interpretation among institutions? How should principal investigators, co-investigators, and research teams be trained in the optimal protection of participants?

Study sponsors, principal investigators, research services, and medical research institutions place tremendous pressures on IRBs. These pressures take many forms, but they are usually aimed at pushing the IRB to approve research studies and to do so in a timely fashion. However, if anything goes wrong in a research study that has been approved by an IRB, the first group to be held accountable by the medical institution is the IRB. Likewise, the principal investigator and study sponsor will say, "The IRB is the embodiment of expertise in the protection of participants at our institution. Why didn't the IRB identify this problem so we could have fixed it so no one would have been hurt? After all, the IRB approved our study." All interested parties must recognize that the way to facilitate research is to provide the best possible protection of study participants and that an institution promotes this goal when it gives its IRB the best possible personnel and enough time to do its job well.

The IRB must be able to assure itself and everyone else that it has all the information necessary to properly review the study. IRB approval should mean:

> This research study and its scientific protocol and informed consent form have been examined and reexamined by the IRB. The IRB assures all that this study offers the best possible protection of study participants and has the aim of generating knowledge for future populations. However, the IRB can offer this assurance only if the relevant research and institutional parties abide by their own assurances.
>
> The study sponsor and principal investigator(s) must be willing to be educated in the protection of study participants and to implement that training in all their research involving human subjects. Further, they must be cognizant and accepting of their responsibility (1) to supply all information to the IRB, especially information about risks and problems, in a timely fashion; (2) to pay constant attention to minimizing risks and to identify problems as early as possible, clearly and precisely communicating about them to all involved parties; and (3) to provide immediate and continuous follow-up of participants to prevent damage and to minimize harm at each step of the research process, from study entry until well after the study has been completed. The results are to be published for scrutiny, so that it can be determined that useful knowledge has been provided, that there has been minimal risk to the participants who bore the risk, and that participants realized that they might not personally benefit (and in some cases might be irreversibly harmed).

> The institution abides by its assurance that it will provide all needed support of the IRB to do its best to protect study participants within the institution. The institution will be mindful of its role of providing a safe haven for individuals to deliberate about participation, to decide to reject or accept a role as study participant, and, no matter what the decision, to be protected by the institution with the highest skill and guidance that the IRB and the institution can provide.

Whether and the extent to which the IRB can achieve its goals depend on all parties' recognizing their roles and obligations in their principal task: the optimal protection of all study participants in all research endeavors.

Appendix 1

A Check List for Reviewing a Scientific Protocol

I. Study sponsor (the individual or group funding the research study)
   A. Is the study sponsor clearly identified?
   B. Is there clarity between the study sponsor and the principal investigator about what is being proposed?
II. Principal investigator(s), co-investigators, and research team
   A. Are all the principal investigators, co-investigators, and members of the research team listed?
   B. Do the principal investigator(s), co-investigators, and research team members collectively have expertise in the following areas?
      1. study participant recruitment
      2. assessment of participants' vulnerability
      3. assessment of participants' decision-making capacity
      4. anonymity, confidentiality, and security of data
      5. physical examination
      6. mental and psychosocial examination
      7. laboratory, test, study performance
      8. laboratory, test, study interpretation
      9. rapid communication to participants and their physicians regarding abnormal findings and adverse outcomes
      10. monitoring for adverse outcomes
      11. prescribing (and renewing) medications
   C. Has the principal investigator reviewed the scientific protocol? If not, why not?
III. Purpose of the study
   A. Does the scientific protocol clearly state the purpose of the study and why humans are needed as participants?
   B. Does the study ask an important medical or scientific question?
   C. What does the study question involve?
      1. the development of more scientific knowledge about:
         a. a medical condition or disease
         b. the diagnosis, management, or treatment of a medical condition or disease
         c. the improvement of health care for persons with a particular medical condition or disease
         d. the determination of the safety or effectiveness of a drug, medical device, or diagnostic or therapeutic procedure
      2. the storage of cells, tissues, organs, or other specimens for use in future research

IV. Type of study
  A. Drug
    1. Does the scientific protocol identify the type of drug study being proposed and state its justification?
      a. single blind
      b. placebo controlled
      c. double blind
      d. drug controlled
      e. open label
      f. cross-over
      g. other (specified)
  B. Device
    1. Does the scientific protocol list any and all risks associated with the use of the device known by the product manufacturer or discoverable in the peer-reviewed medical and scientific literatures?
  C. Radiation in any form
    1. Does the scientific protocol describe the following for each study:
      a. type(s) of radiation to be used (x-ray, CT scan, MRI scan, bone scan, or other)
      b. purpose
      c. frequency of exposure
      d. level of radiation risk
    2. Would each participant have had the same procedure in clinical care, or is the procedure being done for research purposes only?

V. Study design
  A. Is the study design appropriate? For example, in a study involving a new drug:
    1. Is the drug to be compared to placebo only? If so, why?
    2. Is the drug to be compared to the best drug on the market for the particular medical condition or disease? If so, why?

VI. Characteristics of study participants
  A. Who will be included as participants in this study?
    1. volunteers from the community
    2. inpatients (persons hospitalized for medical conditions)
    3. outpatients (persons seen in out-of-hospital clinics)
    4. nursing home residents
    5. hospice patients
    6. students (including but not limited to medical students)
    7. patients' family members
    8. employees
  B. What are the inclusion criteria of the study?
  C. What are the exclusion criteria of the study?
  D. Are both the inclusion and the exclusion criteria fair and just?

1. Are any individuals being inappropriately included as participants?
2. Are any individuals being inappropriately excluded as participants?

E. Who will interview individuals to assess whether they meet the inclusion requirements?

F. Who on the research team will help the interviewer determine whether an individual meets the appropriate criteria?

G. How will the principal investigator and research team determine whether an individual has the decision-making capacity to elect enrollment in the study and whether the individual would, in the perspective of a court, be legally competent to volunteer to participate in a research study?

VII. Vulnerable participants

A. Does the study involve any vulnerable participants?

B. Can the study be done without involving any vulnerable participants? If so, why is any vulnerable participant involved?

C. Who will decide whether a particular vulnerable individual will be allowed to participate in the study?

D. What criteria will be used to decide whether a particular vulnerable individual will be allowed to participate in the study?

E. How will the IRB monitor the participation of vulnerable participants?

VIII. Identification of participants

A. How will individuals be identified for recruitment into the study?

B. Will the institution's computer system be used in the identification process? If so, how?

C. Will recruitment posters or flyers be used? Have they been included in the materials for IRB review?

D. Will patients be recruited from hospital wards or clinics by means other than through the hospital computer system? If so, have all recruitment methods that will be employed been listed?

IX. Study population

A. At how many sites will the study be conducted?

B. How many participants will be involved at each site?

C. Is the study population of the appropriate size to achieve meaningful scientific and statistical results?

X. Involvement of primary care or other clinical care provider(s)

A. If the scientific protocol identifies a primary care or other clinical care provider as having a role in the study (for example, in the recruitment of participants), has that person agreed to participate as described?

XI. Participant involvement

A. Does the scientific protocol state the procedures the participants will undergo during each phase of the study?

B. What will the initial interview process consist of?

1. Will there be a psychological evaluation?

2. Will there be a review of the individual's medical history?
3. Will a physical examination be performed to ensure that the individual is in good health?
4. Will blood testing be done?
    a. If so, what test(s), and why?
    b. How much blood will be drawn?
    c. Who will draw the blood?
    d. Who will obtain informed consent for the blood draw?
    e. When will the participant learn the result(s)?
5. Will any other study(ies) be done?
    a. If so, what study(ies), and what are the risks?
    b. When will the participant learn the result(s)?

C. If the individual is qualified to enroll in the study, how many more visits will be required, over what time interval, and for approximately how long?

D. Will the participant be required to complete a questionnaire?
1. If so, will the questions be personal and/or potentially upsetting?
2. What counseling will be available to individuals during and/or after the questionnaire session(s) to help with their emotional state?

E. Does the study involve taking a placebo?
1. If so, why is a placebo needed?
2. Could the study be done without a placebo?
3. What placebo effects are to be expected?

F. If this is a study of a new drug versus a standard drug or a new drug versus placebo, will participants be randomly assigned to receive the new drug, standard drug, or placebo? If so, what randomization process will be used?

XII. Potential benefits

A. Does the scientific protocol state any potential benefits to an individual volunteering as a participant?

XIII. Potential risks

A. Does the scientific protocol specify all risks (adverse outcomes, their nature, their chance of occurrence, and their impact on a participant's survival and quality of life) to participants during participation in the study?

B. Do the risks stated in the scientific protocol match the risks stated in the informed consent form? If not, why not?

C. Do the risks stated in the scientific protocol match the results of the IRB member's search of the peer-reviewed medical literature or discussion with experts in the field? If not, why not?

XIV. Potential discomforts

A. Does the risk section of the scientific protocol state all discomforts that the participant should be prepared for during and after participation in the study?

XV. If there are particular risks to women (including risks to fetus during pregnancy)

A. Is it known how this treatment could affect an unborn child?

1. If so, is the degree of risk precisely stated in the scientific protocol and informed consent form?
2. If affect not known, how is participant to be informed of this?

B. Is there a requirement that if the female participant is sexually active and capable of becoming pregnant, she must use an effective method of birth control while participating in the study?

XVI. If there are particular risks to men

A. Are there specific risks to men, including risk to sperm from participation in the study or effects of the study drug on sperm?

XVII. Liability

A. Does the scientific protocol specify the following regarding a study participant who sustains an adverse outcome?

1. Who will determine if the adverse outcome is related to participation in the study and what criteria will be used in this determination?
2. Who will define what is and what is not a reasonable and necessary resultant expense for the participant and what criteria will be used in this determination?
3. Who will pay physician and hospital costs?
4. Will compensation be provided for long-term disability resulting from a severe adverse outcome?
5. What legal recourse will a participant have if it is decided that a severe adverse outcome was not related to the study?

B. Does the scientific protocol say how the participant will be informed of each of the above specifications?

XVIII. Alternatives to participation in the study

A. Does the scientific protocol specify the other medications, combinations of medications, and therapies, if any, that are available in standard care for the medical condition or disease being studied?

B. Does it state that the individual considering enrollment may discuss any alternative treatment options with his or her regular physician?

XIX. Participant's health information (including, medical records, history and physical examination records, consultation reports, laboratory test results, x-ray and other diagnostic imaging reports, operative reports, discharge summaries, progress notes, questionnaire, interview results, focus group surveys, psychological surveys, psychological performance test results, photographs, videotapes, and tissues and/or blood specimens)

A. Does the scientific protocol specify which parties may have access to any or all of the participant's health information if authorized by the participant?

B. Does it specify how the principal investigator will determine whether the participant understands how these elements of health information pertain to participation in the study?

C. Does it state how the principal investigator will determine that the participant understands what he or she is authorizing by signing any statement regarding the release of health information collected before and during the study?

D. Does it specify the precautions that the principal investigator and research team will take to ensure that all elements of health information are protected at all times? Are these precautions adequate?

E. Does it specify how the participant will be notified if, despite the best attempts at protection, health information falls into the wrong hands?

XX. Safe participation and withdrawal from the study

A. Does the scientific protocol state the practices and procedures to be followed if a participant elects to withdraw from the study?

B. Does it list all circumstances in which a participant may be asked to discontinue participating in or may be removed from the study? How will the individual be treated safely after discontinuing the study?

C. Does it specify how the principal investigator will convey any new information related to the study which may influence the participant's willingness to continue in the study?

XXI. Costs to the participant, medical or research institution, and others

A. Does the scientific protocol list any and all costs to the participant and his or her insurance company?

B. Does it list any and all costs to the medical or research institution and others?

XXII. Potential conflicts of interest

A. Does the scientific protocol describe in detail any and all potential conflicts of interest of the study sponsor, the medical and research institutions, the principal investigator, and any member of the research team?

XXIII. Storage of research data

A. Where and how will the research data be kept during the study and after its completion? Are the place and the procedures safe and secure?

B. How long will the data be kept?

C. Who will have the primary responsibility for destroying the data?

D. If the data are not to be destroyed, who will be responsible for maintaining their anonymity, confidentiality, and security over time?

E. Is a data safety monitoring board part of the study? If so, where is the board located, who are its members, and how may the IRB, the principal investigator, and the research team communicate with the board?

XXIV. Storage of biologic specimens

A. Will there be storage of biologic specimens for future research?

B. In the case of uniquely identified biologic specimens, especially those containing genetic material, do the participant and his or her family understand where and how their genetic material will be stored and protected and who will have access to it and why?

C. How will this understanding be verified, and what will be done if a participant withholds or withdraws consent for such a donation?

Appendix 2

A Check List for Reviewing an Informed Consent Form

I. Is there a clear distinction made between clinical care and clinical research?
   A. Throughout the descriptions in the form, is there a clear distinction between:
      1. what is being performed not because it is a component of participation in the research study but because it would be performed in clinical care?
      2. what is being performed because of participation in research?
   B. Is the participant's clinician involved in the study?
      1. The fact that a participant's clinician is involved in the study (as a principal investigator, co-investigator, or clinician expert reviewing laboratory tests or studies) can be problematic. It may, over time, cause the participant to fail to distinguish between clinical care and those procedures that are due solely to participation in the research study.
   C. Is the setting where individuals are being recruited or would report for research-related activities the same as where they are seen for clinical care?
      1. If so, this may cause confusion about what is research activity and what standard care.

II. Are the terms participants will encounter in research clearly defined?
   A. Is there a clear explanation of the terms *medical research, clinical trial,* and *clinical investigation,* which have similar meanings?
      1. If the terms are used equivalently, then this should be pointed out.
   B. Are other research terms, such as *randomization,* clearly defined and illustrated?
      1. If verbal probability terms, such as *rare,* are used, are corresponding best estimated numerical terms also provided?
      2. Is the fact that no benefit will necessarily accrue to the participant made clear?
   C. Is the principal investigator providing the best estimates of the risks involved?
   D. Are the costs to the study participant made clear?
      1. time required for study participation
      2. costs to self and others for transportation to and from the research site
      3. time away from work and family
      4. emotional upset from questionnaire studies
      5. that social stigmatization may occur if data protection fails
   E. Are the alternatives for care that exist outside the study stated clearly?

III. Does the form describe how study records will be kept secure, to protect the privacy of the patient's medical record?
   A. Will separate records be kept for the study, or will some of the research records be made part of the patient's medical records?
      1. Is it clear to whom the participant would be granting permission to see his or her medical records or research records?

IV. Have the participant's rights of withdrawal and redress of grievances been clearly stated?
   A. Is it clear that the participant has the right to withdraw from the study at any time, for whatever reason, within the bounds of safety?
   B. Is it clear that the participant has a right of suit for study-related injuries?
   C. Will it be stated in the patient's medical record that he or she is a participant in the study (or that he or she withdrew from the study)?
      1. Is it clear that the participant's clinical care provider will be informed of key findings (for example, abnormalities detected in the study)?
      2. How will the IRB protect the patient's confidentiality in a study involving, for example, an illicit activity (for example, a participant's use of illegal drugs or other substances)?
      3. Is there a clear specification of how data will be kept private, to what extent the protection is in place, and what will happen if, despite the safeguards, data are disclosed?

V. Have all institutional policies and state and local laws been considered (e.g., certain states have genetic privacy laws that need to be followed)?

Notes

The portions of *The Code of Federal Regulations* referred to below are Title 38 ("Pensions, Bonuses, and Veterans' Relief"), Chapter 1 ("Department of Veterans Affairs"), Part 16 ("Protection of Human Subjects"), cited below as 38 *CFR* 16, and Title 45 ("Public Welfare"), Subtitle A ("Department of Health and Human Services"), Part 46 ("Protection of Human Subjects"), cited as 45 *CFR* 46.

Preface

1. C. Weijer and P. B. Miller, "Therapeutic Obligation in Clinical Research," *Hastings Center Report* 33, no. 3 (May–June 2003): 3.

1 / What Is an IRB, and What Does It Do?

1. 38 *CFR* 16 and 45 *CFR* 46.
2. 38 *CFR* 16.107.a and 45 *CFR* 46.107.a.
3. Ibid.
4. Ibid.
5. 45 *CFR* 46.111.
6. 38 *CFR*.
7. 38 *CFR* 16.107 and 45 *CFR* 46.107.
8. 38 *CFR* 16.108.b and 45 *CFR* 46.108.b.
9. 38 *CFR* 16.107.c and 45 *CFR* 46.107.a.
10. 38 *CFR* 16.107.d and 45 *CFR* 46.107.a.
11. 38 *CFR* 16.107.e and 45 *CFR* 46.107.e.
12. 38 *CFR* 16.107.f and 45 *CFR* 46.107.f.
13. 38 *CFR* 16.101.a and 45 *CFR* 46.101.a.
14. 38 *CFR* 16.115.a and 45 *CFR* 46.115.a.
15. 38 *CFR* 16.115.a.1 and 45 *CFR* 46.115.a.1.
16. 38 *CFR* 16.115.a.2 and 45 *CFR* 46.115.a.2.
17. 38 *CFR* 16.115.a.3 and 45 *CFR* 46.115.a.3.
18. 38 *CFR* 16.115.b and 45 *CFR* 46.115.b.

2 / Basic Terms and Concepts Used in IRB Work

1. D. J. Mazur, "The Problems with the Development of a Consistent Set of Medical-Legal Criteria and Operational Definitions for Use in Regulatory and Tort Law" (Ph.D. diss., Stanford University, 1983).

2. National Commission for the Protection of Human Subjects of Biomedical and Behavioral Research, *The Belmont Report: Ethical Guidelines for the Protection of Human*

Subjects of Research, DHEW Publication o.s. 78-0012. Washington, D.C., 1978. Accessible at http://ohsr.od.nih.gov/guidelines/belmont.html.

3. 38 *CFR* 16.102.d and 45 *CFR* 46.102.d.

4. Ibid.; *Belmont Report,* pp. 2–4.

5. J. Sugarman, N. E. Kass, S. N. Goodman, P. Perentesis, P. Fernandes, and R. R. Faden, "What Patients Say about Medical Research," *IRB* 20, no. 4 (July–August 1998): 1–7. N. E. Kass and J. Sugarman, "Are Research Subjects Adequately Protected? A Review and Discussion of Studies Conducted by the Advisory Committee on Human Radiation Experiments," *Kennedy Institute of Ethics Journal* 6, no. 3 (September 1996): 271–82. N. E. Kass, J. Sugarman, R. Faden, and M. Schoch-Spana, "Trust: The Fragile Foundation of Contemporary Biomedical Research," *Hastings Center Report* 26, no. 5 (September–October 1996): 25–29.

6. *Belmont Report,* p. 3; 38 *CFR* 16.102.d and 45 *CFR* 46.102.d.

7. Ibid.

8. Ibid.

9. Ibid., pp. 4–20.

10. Ibid., p. 3.

11. Ibid.

12. Ibid., p. 4.

13. Ibid., p. 1.

14. 38 *CFR* 16.116.a.5 and 45 *CFR* 46.116.a.5.

3 / What Is Risk?

1. P. Slovic, *The Perception of Risk* (Sterling, Va.: Earthscan Publications, 2000). N. Pidgeon, R. E. Kasperson, and P. Slovic, eds., *The Social Amplification of Risk* (New York: Cambridge University Press, 2003). J. Flynn, P. Slovic, and H. Kunreuther, eds., *Risk, Media, and Stigma: Understanding Public Challenges to Modern Science and Technology* (Sterling, Va.: Earthscan Publications, 2001).

2. D. J. Mazur and D. H. Hickam, "Patients' Preferences for Risk Disclosure and Role in Decision Making for Invasive Medical Procedures," *Journal of General Internal Medicine* 12, no. 2 (February 1997): 114–17.

3. D. Kahneman, P. Slovic, and A. Tversky, *Judgment under Uncertainty: Heuristics and Biases* (New York: Cambridge University Press, 1982). A. J. Lloyd, "The Extent of Patients' Understanding of the Risk of Treatments," *Quality in Health Care* 10, Suppl. 1 (September 2001): i14–18.

4. D. Kahneman and A. Tversky, "On the Reality of Cognitive Illusions," *Psychological Review* 103, no. 3 (July 1996): 582–91; discussion, 592–96. Kahneman, Slovic, and Tversky, *Judgment under Uncertainty.* P. B. Vranas, "Gigerenzer's Normative Critique of Kahneman and Tversky," *Cognition* 76, no. 3 (September 2000): 179–93.

5. Kahneman, Slovic, and Tversky, *Judgment under Uncertainty.*

6. Mazur and Hickam, "Patients' Preferences for Risk Disclosure and Role in Decision Making for Invasive Medical Procedures."

7. D. J. Mazur and D. H. Hickam, "Patient Interpretations of Terms Connoting Low Probabilities When Communicating about Surgical Risk," *Theoretical Surgery* no. 8 (1993): 143–45. D. J. Mazur and J. F. Merz, "Patients' Interpretations of Verbal Expressions of Probability: Implications for Securing Informed Consent to Medical Interventions," *Behavioral Sciences and the Law* 12, no. 4 (Autumn 1994): 417–26. D. J. Mazur and J. F. Merz, "How Age, Outcome Severity, and Scale Influence General Medicine Clinic Patients' Interpretations of Verbal Probability Terms," *Journal of General Internal Medicine* 9, no. 5 (May 1994): 268–71. D. J. Mazur and D. H. Hickam, "Patients' Interpretations of Probability Terms," *Journal of General Internal Medicine* 6, no. 3 (May–June 1991): 237–40.

8. 38 *CFR* 16.116 and 45 *CFR* 46.116.

9. Ibid.

4 / Prescreening of Proposals

1. 38 *CFR* 16.116 and 45 *CFR* 46.116.

5 / The Scientific Protocol

1. 38 *CFR* 16.102.d and 45 *CFR* 46.102.d.

2. 38 *CFR* 16.107.a and 16.111.b and 45 *CFR* 46.107.a and 46.111.b.

3. 38 *CFR* 16.111.b and 45 *CFR* 16.111.b.

4. Ibid.

5. 38 *CFR* 16.116.a.2 and 45 *CFR* 16.116.a.2.

6 / The Informed Consent Form

1. See, e.g., I. Krass, B. L. Svarstad, and D. Bultman, "Using Alternative Methodologies for Evaluating Patient Medication Leaflets," *Patient Education and Counseling* 47, no. 1 (May 2002): 29–35.

8 / Research involving Questionnaires and Surveys

1. D. O. Sears, L. A. Peplau, and S. E. Taylor, *Social Psychology,* 7th ed. (Englewood Cliffs, N.J.: Prentice Hall, 1991), p. 31.

2. Ibid.

3. 38 *CFR* 16.116.a.1 and 45 *CFR* 46.116.a.1.

9 / Protection of Participants' Privacy in Research Data and Specimens

1. Office of Civil Rights, Department of Health and Human Services, "Standards for Privacy of Individually Identifiable Health Information: Final Rules," *Federal Register* 67, no. 157 (August 14, 2002): 53182–273.

2. 45 *CFR* 164.514.b.

3. Ibid.

4. 45 *CFR* 164.514.a.

5. B. Malin, L. Sweeney, "A Secure Protocol to Distribute Unlinkable Health Data,"

AMIA Annual Symposium Proceedings 2005, pp. 485–89. B. Malin and L. Sweeney, "How (Not) to Protect Genomic Data Privacy in a Distributed Network: Using Trail Re-identification to Evaluate and Design Anonymity Protection Systems," *Journal of Biomedical Informatics* 37, no. 3 (June 2004): 179–92. L. Sweeney, "Weaving Technology and Policy together to Maintain Confidentiality," *Journal of Law and Medical Ethics* 25, nos. 2–3 (Summer–Fall 1997): 98–110, 82. L. Sweeney, "Guaranteeing Anonymity When Sharing Medical Data: The Datafly System," *Proceedings of the AMIA Annual Fall Symposium 1997*, pp. 51–55. L. Sweeney, "Replacing Personally-Identifying Information in Medical Records: The Scrub System," *Proceedings of the AMIA Annual Fall Symposium 1996*, pp. 333–37.

10 / The Ethical Issues of Informed Consent

1. *Slater v. Baker and Stapleton*, 95 Eng. Rep. 860; 2 Wils. K.B. 359 (1767).

2. *Schloendorff v. Society of New York Hospitals*, 211 N.Y. 125; 105 N.E. 92 (1914).

3. *Salgo v. Leland Stanford Junior University Board of Trustees*, 154 Cal. App. 2d 560; 317 P.2d 170 (1957).

4. *Canterbury v. Spence*, 464 F.2d 772 (1972).

5. R. R. Faden and T. L Beauchamp, *A History and Theory of Informed Consent* (New York: Oxford University Press, 1986).

6. P. S. Appelbaum, "Rethinking the Conduct of Psychiatric Research," *Archives of General Psychiatry* 54, no. 2 (February 1997): 117–20. P. S. Appelbaum and T. Grisso, "The MacArthur Treatment Competence Study, I: Mental Illness and Competence to Consent to Treatment," *Law and Human Behavior* 19, no. 2 (April 1995): 105–26. J. W. Berg, P. S. Appelbaum, and T. Grisso, "Constructing Competence: Formulating Standards of Legal Competence to Make Medical Decisions," *Rutgers Law Review* 48, no. 2 (Winter 1996): 345–71. D. J. Moser, S. K. Schultz, S. Arndt, M. L. Benjamin, F. W. Fleming, C. S. Brems, J. S. Paulsen, P. S. Appelbaum, and N. C. Andreasen, "Capacity to Provide Informed Consent for Participation in Schizophrenia and HIV Research," *American Journal of Psychiatry* 159, no. 7 (July 2002): 1201–7.

7. National Commission for the Protection of Human Subjects of Biomedical and Behavioral Research, *The Belmont Report: Ethical Principles and Guidelines for the Protection of Human Subjects of Research.* DHEW Publication, o.s., 78-0012. Washington, D.C., 1978. Accessible at http://ohsr.od.nih.gov/guidelines/belmont.html.

8. Ibid., p. 11.

9. Ibid., pp. 11, 14–15.

11 / Continuing Review, Communication, and Feedback

1. 38 *CFR* 16.113 and 45 *CFR* 46.113.

14 / Decision-Making Capacity and Accountability in Research

1. G. Szmukler, "Double Standard on Capacity and Consent?" *American Journal of Psychiatry* 158, no. 1 (January 2001): 148–49.

2. B. C. White, *Competence to Consent* (Washington, D.C.: Georgetown University Press, 1994), p. 45.

3. Ibid., p. 74.

4. U.S. President's Commission for the Study of Ethical Problems in Medicine and Biomedical and Behavioral Research, *Making Health Care Decisions: A Report on the Ethical and Legal Implications of Informed Consent in the Patient-Practitioner Relationship,* vol. 1 (Washington, D.C., 1982), p. 57.

5. C. Elliott, "Caring about Risks: Are Severely Depressed Patients Competent to Consent to Research?" *Archives of General Psychiatry* 54, no. 2 (February 1997): 113–16.

6. T. Grisso and P. S. Appelbaum, "The MacArthur Treatment Competence Study, III: Abilities of Patients to Consent to Psychiatric and Medical Treatments," *Law and Human Behavior* 19, no. 2 (April 1995): 149–74.

7. P. S. Appelbaum, "Rethinking the Conduct of Psychiatric Research," *Archives of General Psychiatry* 54, no. 2 (February 1997): 117–20.

8. R. J. Bonnie, "Research with Cognitively Impaired Subjects: Unfinished Business in the Regulation of Human Research," *Archives of General Psychiatry* 54, no. 2 (February 1997): 105–11.

9. Elliott, "Caring about Risks," p. 113, 114.

10. R. R. Faden and T. L Beauchamp, *A History and Theory of Informed Consent* (New York: Oxford University Press, 1986), pp. 252 and 300.

11. T. Grisso and P. S. Appelbaum, "Comparison of Standards for Assessing Patients' Capacities to Make Treatment Decisions," *American Journal of Psychiatry* 152, no. 7 (July 1995): 1033–37.

12. Elliott, "Caring about Risks," p. 114. Italics in original.

13. Ibid., p. 115. Italics in original.

14. Ibid.

15. Ibid., p. 116.

Summary / The IRB's Key Role

1. T. L. Beauchamp and J. F. Childress, *Principles of Biomedical Ethics,* 5th ed. (New York: Oxford University Press, 2001).

Website References for Cited Documents

Belmont Report (1979)
http://ohsr.od.nih.gov/guidelines/belmont.html

Code of Federal Regulations (Department of Health and Human Services, National Institutes of Health, Office for Protection from Research Risks): Title 38 ("Pensions, Bonuses, and Veterans' Relief"), Chapter 1 ("Department of Veterans Affairs"), Part 16 ("Protection of Human Subjects"), revised as of July 1, 2005; Title 45 ("Public Welfare"), Subtitle A ("Department of Health and Human Services"), Part 46 ("Protection of Human Subjects"), revised as of October 4, 2004
www.gpoaccess.gov/cfr/

Declaration of Helsinki ("Code, Amendments, and Clarifications, 1964–2000")
www.wma.net/e/policy/b3.htm

HIPAA Privacy Regulations (1996)
www.hhs.gov/ocr/hipaa

Interpreting the Common Rule for the Protection of Human Subjects
www.nsf.gov/bfa/dias/policy/hsfaqs.jsp

Index